H.-J. Lenz   G. B. Wetherill
P.-Th. Wilrich (Eds.)

Frontiers in Statistical Quality Control 4

With 57 Figures

Springer-Verlag Berlin Heidelberg GmbH

Professor Dr. Hans-Joachim Lenz, Institut für Statistik und Ökonometrie,
Freie Universität Berlin, Garystraße 21
W-1000 Berlin 33, FRG

Professor G. Barry Wetherill, The University of Newcastle upon Tyne,
Industrial Statistics Research Unit, GB-Newcastle upon Tyne NE1 7RU
England

Professor Dr.-Ing. Peter-Theodor Wilrich, Institut für Statistik und
Ökonometrie, Freie Universität Berlin, Garystraße 21
W-1000 Berlin 33, FRG

ISBN 978-3-7908-0642-7   ISBN 978-3-662-11789-7 (eBook)
DOI 10.1007/978-3-662-11789-7

© Springer-Verlag Berlin Heidelberg 1992
Originally published by Physica-Verlag Heidelberg in 1992

Frontiers in
Statistical Quality Control 4

Editorial

The 4th International Workshop on Statistical Quality Control took place in 1990. Our host was the Louisiana State University at Baton Rouge, Louisiana, USA. We are deeply indebted to Professor Helmut **Schneider**, of the Louisiana State University, for making all the practical arrangements.

This occasion was the first time that the Workshop has been held outside Europe. There were 30 participants from the USA, Europe, Japan and Israel. We now feel that this is about the right number of participants for the Workshop. We always find that many fruitful discussions occur aside from the organised programme, and a group of any larger than 30 would reduce the Workshop to a 'Conference'. Indeed, some of the informal activities organised by Professor **Schneider** and his staff were happy and fruitful occasions, bringing collaboration between workers who live in distant parts of the world.

The papers presented in this volume fall into three groups. In the first group there are nine papers on acceptance sampling. The first paper is by **Baillie**, on a method of combining attributes and variables plans, which he calls attriables. **Koyama** studies the sensitivity to process quality when operating under ISO 2859. **Kaijage**, **Hall** and **Hassan** propose a sampling plan when the prior distribution of lot defectives is of a non-binomial type. **Kimura** and **Asano** introduce a software package for studying the properties of sampling inspection plans. **Bruhn-Suhr**, **Krumbholz** and **Lenz** present an algorithm for sampling inspection by variables using an (approximate) ML estimator of the fraction detective. **Von Collani** deals with some contradictions in the propositions and the decision rules of acceptance sampling plans by variables. Economic sampling plans are analysed by **Ohta** and **Ogawa** using an empirical Bayesian approach and by **Seidel** using the minimax principle linked to the moments of the prior distribution.

The second large group of papers deal with control charts and process control. **Hart** and **Hart** reconsider single-value charts. **Thyregod** and **Madsen** present a control chart for auto-correlated processes using Kalman filtering. **Adams**, **Lowry** and **Woodall** discuss the misinterpretation and misuse of false alarm probabilities in the chart design. **Ladany** and **Raz** present work on a dynamic programming model for a control chart. **Shahani** shows the need for a better understanding of the process dynamics and the control actions. **Saniga** studies economic designs with constraints. **Hryniewicz** proposes simplified process control procedures using an approximate economic model. **Schneider**, **Hui** and **Pruett** deal with an interesting application of control charts to environmental data. **Wetherill** and **Brown** discuss the special problems which arise in the process industries.

The third group of papers include contributions by **Benski**, and by **Riebschläger** and **Wilrich** on experimental design with special emphasis on unreplicated factorial designs and orthogonal factorial designs.

As we look forward to the 5th Workshop, we would like to see some papers on software systems for quality control, some contributions by industrial practitioners, and more work related to process industry problems and experimental design in addition to the standard coverage of acceptance sampling and control charts.

Berlin, Newcastle upon Tyne   The Editors

Contents

PART 3: EXPERIMENTAL DESIGN

Author Index

Adams, B.M., Prof., Dept. of Management Science and Statistics,
    The University of Alabama, 75 Bidgood, Box 870226, Tuscaloosa,
    AL 35487-0226, USA
Asano, Ch., Prof., Dept. of Information System Science, Faculty of
    Engineering, Soka University, Hachiouji, Tokyo, 192 Japan
Baillie, D.H., Dr., Statistical Concepts Branch, Procurement
    Executive, UK Ministry of Defence, St. Christopher House,
    Southwark Street, London SE1 OTD, United Kingdom
Benski, C., Dr., Merlin Gerin, Centre de Recherches - A2,
    F-38050 Grenoble Cedex, France
Brown, D.W., Dr., ICI Chemicals & Polymers Group, P.O. Box 90,
    Wilton, Middlesbrough, Cleveland TS6 8JE, United Kingdom
Bruhn-Suhr, M., Dr., Universität Hamburg, Modellversuch
    Fernstudienzentrum, Averhoffstr. 38, D-2000 Hamburg 76,
    Germany
Collani, E.v., Prof., Institut für Angewandte Mathematik und
    Statistik, Universität Würzburg, Sanderring 2,
    D-8700 Würzburg, Germany
Hall, J.E., Dr., Dept. of Information and Decision Sciences,
    University of Illinois at Chicago, Chicago, Illinois 60680,
    USA
Hart, M.K., Prof., College of Business Administration, The
    University of Wisconsin Oshkosh, Oshkosh, Wisconsin 54901, USA
Hart, R.F., Prof., College of Business Administration, The
    University of Wisconsin Oshkosh, Oshkosh, Wisconsin 54901, USA
Hassan, M.Z., Prof., Stuart School of Business, Illinois
    Institute of Technology, 10 West 31st Street, Chicago,
    Illinois 60616, USA
Hryniewicz, O., Prof., Systems Research Institute, Polish Academy
    of Sciences, Newelska 6, 01-447 Warsaw, Poland
Hui, Y., Mrs., Dept. of Quant. Business Analysis, Louisiana State
    University, Baton Rouge, LA 70803, USA
Kaijage, E.L., Dr., Soft Sheen Products, Inc., 1000 East 87th
    Street, Chicago, Illinois 60619
Kimura, H., Prof., Dept. of Applied Mathematics, Faculty of
    Science, Okayama University of Science, Ridaicho, Okayama,
    700 Japan
Koyama, T., Prof., Faculty of Engineering, Tokushima Bunri
    University, Shido-cho, Kagawa, Japan 769-21
Krumbholz, W., Prof., Universität der Bundeswehr Hamburg, Postfach
    70 08 22, Holstenhofweg 85, D-2000 Hamburg 70, Germany
Ladany, S.P., Prof., Dept. of Industrial Engineering and
    Management, Ben-Gurion University of the Negev, P.O. Box 653,
    Beer Sheva 84105, Israel
Lenz, H.-J., Prof., Institut f. Statistik und Ökonometrie, Freie
    Universität Berlin, Garystr. 21, D-1000 Berlin 33, Germany
Lowry, C.A., Prof., Finance and Decision Science Department,
    M.J. Neeley School of Business, Texas Christian University,
    P.O. Box 32868, Fort Worth, TX 76129, USA
Madsen, H., Prof., IMSOR, Bygning 321, The Technical University
    of Denmark, 2800 Lyngby, Denmark

Contents

Part 1
Sampling Inspection

Attriables Acceptance Sampling Plans

D.H. Baillie, London, England

1. INTRODUCTION

Consider a class or group of quality characteristics which are of approximately equal criticality to the acceptability of a product, process or service, and to which a single overall acceptable quality level (AQL) applies, e.g. Major, Minor A, Minor B etc. A self-evidently desirable property of any multivariate acceptance sampling procedure for such a class is that the probability of acceptance when the process fraction nonconforming of the class is p should not depend upon the way in which the individual quality characteristics in the class contribute to p. Baillie (1987) for brevity introduced this property as "partition-invariance" and developed approximately partition-invariant single sampling inspection procedures for two or more normally distributed variables with unknown means. The key idea underlying those developments was to base the acceptance criterion on the minimum variance unbiased (MVU) estimator of p, as had been done for the corresponding univariate cases by Bowker and Goode (1952) and Lieberman and Resnikoff (1955).

Two methods presently prevail for combining attributes and variables data from different quality characteristics for acceptance sampling purposes. One is to convert the variables into attributes and to use an attributes acceptance criterion. This method does have the advantage of being either approximately or exactly partition-invariant, but it is highly wasteful of variables information that is often expensive to acquire; as a result, the power of discrimination between good and bad process quality is reduced. The other method, given in ISO 3951 : 1989 is effectively to have a "super-plan" consisting of one multiattribute plan together with a small number of individual variables plans, the super-plan being to accept the inspection lot if and only if each of the individual plans would have led to an accept decision. The disadvantages of such an approach for quality characteristics of equal criticality are two-fold. Firstly, when at least one of the quality characteristics is a variable there is no automatic compensation between deterioration in one quality characteristic and a counterbalancing improvement in another, which poses a delicate problem of how to choose the individual plans. Secondly, no matter how these individual plans are chosen, the super-plan will be far from partition-invariant. For example, in their recent paper that was based on this type of approach, Danziger and Papp (1988) observed:

> "The dependence of the OC-band on the plan is complex and, at the present time, we
> do not have a direct mathematical relation that would assure a plan having optimum char-
> acteristics such as a narrow range of acceptance probabilities and risks less than or equal
> to α or β." (The risks α and β are the producer's and consumer's risks respectively.)

In the present paper, acceptance sampling plans based on the MVU estimator of p are developed for the case where the quality characteristics are of approximately equal criticality and include at least one attribute and at least one variable, with the variables normally distributed. These plans are termed "attriables" plans to distinguish them from, for example, the combined variables and attributes plans of Bowker and Goode (1952) and the mixed variables and attributes plans of Gregory and Resnikoff (1955), all of which are essentially double sampling plans applied to a single quality characteristic, treating it as a variable in the first sample and as an attribute in the first and second sample combined.

We begin by considering acceptance inspection of one attribute in combination with one normally distributed variable of unknown mean μ and unknown variance σ^2, first for the case of independence between the attribute and the variable and then for one particular form of dependence. Implications concerning how variables and attributes plans should be matched are discussed, plans matching those of ISO 2859-1 : 1989 are derived and their high degree of partition-invariance demonstrated by examples. Generalizations to the case of one or more attributes in combination with one or more variables are described. Finally, plans are developed for the case where the variables have known variances and covariances.

2. ONE ATTRIBUTE AND ONE INDEPENDENT VARIABLE

2.1 The Overall Process Fraction Nonconforming, p, under Independence

Consider the combined inspection of one normally distributed random attribute and one independently distributed random variable. Denote the process fraction nonconforming for the variable by p_v, for the attribute by p_a and overall by p. From the independence of the variable and the attribute, it follows that the overall process fraction *conforming* equals the product of the process fractions conforming for the variable and the attribute, i.e. $1-p = (1-p_v)(1-p_a)$ which on rearranging gives

$$p = 1 - (1-p_v)(1-p_a). \tag{1}$$

2.2 The MVU Estimator of p under Independence

Consider the case where neither the mean, μ, nor the variance, σ^2, of the variable are known. Under these circumstances the standard acceptance sampling procedure for a single variable is called the $"s"$ method, so the subscript s will be used to indicate ignorance of the process mean(s) and (co)variance(s).

Suppose that from each inspection lot a random sample of n_s items is drawn for the purpose of inspecting both the attribute and the variable and a further random sample of $n_a - n_s$ $(n_a \geq n_s)$ items is drawn for the purpose of inspecting the attribute only. Suppose further that the lower and upper combined double specification limits on the variable are L and U respectively. Denote the sample values on the variable by $x_1, x_2, ..., x_{n_s}$ and the number of nonconforming items in the combined samples by r. Let

$$\bar{x} = \sum_{i=1}^{n_s} x_i / n_s \tag{2}$$

and

$$S^2 = \sum_{i=1}^{n_s} (x_i - \bar{x})^2. \tag{3}$$

Finally. denote the unique uniformly MVU estimators of p_v, p_a and p by $\hat{p}_s$, $\hat{p}_a$ and $\hat{p}$ respectively. Then Appendix A shows that $\hat{p}_s$, $\hat{p}_a$ and $\hat{p}$ are related in the same way as p_v, p_a and p, i.e.

$$\hat{p} = 1 - (1-\hat{p}_s)(1-\hat{p}_a) \tag{4}$$

with

$$\hat{p}_a = r / n_a \tag{5}$$

and

$$\hat{p}_s = B(\tfrac{1}{2}\max[0, \ 1-\{(U-\bar{x})\sqrt{n_s}\}/\{S\sqrt{(n_s-1)}\}])$$

$$+ \ B(\tfrac{1}{2}\max[0, \ 1-\{(\bar{x}-L)\sqrt{n_s}\}/\{S\sqrt{(n_s-1)}\}]) \tag{6}$$

where $B(.)$ denotes the distribution function of the symmetric beta distribution of the first kind with parameters $\tfrac{1}{2}n_s-1$. The MVU estimator of p under independence for a single specification limit follows immediately by setting L to $-\infty$ or U to ∞.

2.3 Matching Variables and Attributes Plans

2.3.1 Equality of variables and attributes acceptance constants

The acceptance criteria investigated in this paper are all of the same generic form, i.e. accept if and only if $\hat{p} \le p^*$, where p^* is a suitably chosen acceptance constant. For the case of one random normal variate in combination with one independent random attribute, $\hat{p}$ is given by (4) in conjunction with (5) and (6), but later in this paper more general forms for $\hat{p}$ will be considered.

By definition, for the resulting plans to be approximately partition-invariant, it is necessary for the probability of acceptance of an inspection lot at a given process fraction nonconforming, p, to be roughly constant regardless of how p is apportioned between the variable and the attribute, for all p. Two extreme cases are when $(p_v, p_a) = (p, 0)$ and when $(p_v, p_a) = (0, p)$. Thus we require

$$P(\hat{p} \le p^* \mid p_v = p, \ p_a = 0) \approx P(\hat{p} \le p^* \mid p_v = 0, \ p_a = p) \quad \text{for } 0 \le p \le 1,$$

i.e.

$$P(\hat{p}_s \le p^* \,|\, p_v = p) \approx P(\hat{p}_a \le p^* \,|\, p_a = p) \quad \text{for } 0 \le p \le 1. \tag{7}$$

But (7) implies that the variables plan and the corresponding attributes plan must have approximately coincident operating characteristic (OC) curves, *whilst at the same time having the same acceptance constant!*

2.3.2 Direction of matching, and constraint on the value of the acceptance constant

It remains to decide how to match variables and attributes plans subject to this constraint. The first decision to be made is whether to match attributes plans to variables plans or vice versa. Consider the consequences of matching an attributes plan to a variables plan. The variables plan is defined by the sample size and acceptance constant (n_s, p^*), so p^* would already be fixed by the variables plan. Optimising the correspondence, however defined, between the attributes and variables OC curves could therefore only be carried out with respect to the attributes sample size, n_a. Note that the acceptance criterion for the attributes plan is

accept if and only if $\hat{p} = r/n_a \le p^*$,

i.e.

accept if and only if $r \le n_a p^*$

from which it follows that the acceptance number of the attributes plan is the integral part of $n_a p^*$.

Now consider matching in the opposite direction, i.e. variables to attributes. The attributes plan would be defined by a sample size n_a and an acceptance number, say Ac. Acceptance would take place if $r \le Ac$. As we have just seen, Ac is the integral part of $n_a p^*$, so $Ac \le n_a p^* < Ac + 1$. Hence

$$Ac/n_a \le p^* < (Ac + 1)/n_a. \tag{8}$$

Optimising the correspondence between the variables and attributes OC curves may therefore be carried out with respect to both n_s and p^*, the latter subject to the constraint (8). One might reasonably expect this bivariate optimisation to tend to produce a closer correspondence between the OC curves than the univariate optimisation implicit in matching attributes plans to variables plans.

Another good reason for matching variables plans to attributes plans rather than vice versa is that, because of the absence of suitable combined attributes and variables plans, many users work entirely in terms of attributes, so there may not be a variables plan available against which to match. A third reason is that attributes plans have been around longer, and are much better established than variables plans in most acceptance inspection environments. It was therefore decided to match variables plans to standard attributes plans.

2.3.3 Matching constraint on variables sample sizes

When the variables sample sizes are not constant across the range of acceptable quality levels (AQLs) for a given lot or batch size, the implementation of multi-class plans becomes complicated, with different numbers of observations being required on different variables. Even in the univariate case this problem arises in the separate and complex double specification limit cases. It was therefore decided to constrain the variables sample size to be constant across AQLs for a given lot size, so that this problem did not arise.

2.3.4 Method of matching

One method of matching a variables plan to an attributes plan is to choose the sample size and acceptance constant of the variables plan to minimise the sum of the absolute vertical discrepancies between the OC curves of the two plans at two given qualities. This method was used in a generalised form, the variables plans matching the k attributes plans defined by (n_a, Ac_1), (n_a, Ac_2), ..., (n_a, Ac_k) being found by minimizing

$$\sum_{i=1}^{k}\sum_{j=1}^{2} |Pv(p_{ij}|n_s, p_i^*) - Pa(p_{ij}|n_a, Ac_i)| \tag{9}$$

with respect to n_s, p_1^*, p_2^*, ..., p_k^* and subject to (8), where Pa and Pv denote the probability of acceptance for attributes and variables respectively and where p_{i1} and p_{i2} are given by $Pa(p_{i1}|n_a, Ac_i) = 0.90$ and $Pa(p_{i2}|n_a, Ac_i) = 0.50$ for $i = 1, 2, ..., k$. The choice of p_{i1} and p_{i2} as the process fractions nonconforming under sampling by attributes corresponding to probabilities of acceptance of 0.90 and 0.50 concentrated the matching on the upper portion of the OC curves.

Tables 1 and 2 show variables plans for lot-by-lot inspection which respectively match the normal and tightened inspection plans of ISO 2859-1 : 1989. For a given sample size code letter the parameter k in (9) represents the number of distinct normal and tightened inspection plans with AQLs in the range 0.1% to 10%.

For sample size code letter A the variables sample size is 3 whilst the attributes sample size is 2, so the inspection procedure given in section 2.2 needs some modification. For code letter A under independence, select a random sample of $n_a = 2$ items from the lot for the purpose of inspecting both the attribute and the variable, and a further random sample of $n_s - n_a = 1$ item for the purpose of inspecting the variable only. Then proceed as before.

For a selection of combined plans from Table 1, Figure 1 shows just how good the matching is between attributes OC curves and the corresponding "s" method variables OC curves for a single specification limit. In general, the constant variables sample size constraint causes the poorest fits to occur at the lower AQLs, where the variables OC curve is too discriminating.

2.3.5 Reduced inspection plans

The matching of variables plans to reduced inspection attributes plans was carried out as

Table 1. Single Sampling Attriables ("s" Method) Plans for Normal Inspection (Master Table)

Sample size code letter	Attributes sample size, n_a	Variables sample size, n_s	0.10	0.15	0.25	0.40	0.65	1.0	1.5	2.5	4.0	6.5	10.0
						Acceptable Quality Levels (Normal Inspection) in % — $10^5 p*$							
A	2	3	↓	↓	↓	↓	↓	↓	↓	↓	↓	31897	↓
B	3	3	↓	↓	↓	↓	↓	↓	↓	↓	↓	19524	↑
C	5	4	↓	↓	↓	↓	↓	↓	↓	8346	↑	↓	32362
D	8	6	↓	↓	↓	↓	↓	↓	5297	↑	↓	19505	32303
E	13	9	↓	↓	↓	↓	↓	3344	↑	↓	11456	19590	27457[3]
F	20	13	↓	↓	↓	↓	2273	↑	↓	7250	12454	17633	27800[1]
G	32	20	↓	↓	↓	1520	↑	↓	4502	7660	10857	17244	23578[1]
H	50	31	↓	↓	1042	↑	↓	2912	4889	6899	10950	15005[5]	21066[1]
J	80	48	↓	688	↑	↓	1847	3063	4302	6806	9324	13107[3]	18145[1]
K	125	70	457	↑	↓	1196	1969	2754	4342	5943	8352	11571[2]	17202[1]
L	200	98	↑	↓	754	1234	1721	2706	3700	5198	7202	10716[1]	↑
M	315	139	↓	485	789	1097	1720	2348	3294	4562	6786[5]	↑	↑
N	500	181	307	498	691	1082	1476	2070	2865	4262	↑	↑	↑
P	800	232	312	432	676	921	1291	1786	2656	↑	↑	↑	↑
Q	1250	286	277	432	589	824	1140	1696	↑	↑	↑	↑	↑
R	2000	351	270	367	514	711	1057	↑	↑	↑	↑	↑	↑

Table 2. Single Sampling Attriables ("s" Method) Plans for Tightened Inspection (Master Table)

Sample size code letter	Attributes sample size, n_a	Variables sample size, n_s	0.10	0.15	0.25	0.40	0.65	1.0	1.5	2.5	4.0	6.5	10.0
						Acceptable Quality Levels (Normal Inspection) in % — $10^5 p*$							
A	2	3	↓	↓	↓	↓	↓	↓	↓	↓	↓	↓	↑
B	3	3	↓	↓	↓	↓	↓	↓	↓	↓	↓	19524	↑
C	5	4	↓	↓	↓	↓	↓	↓	↓	8436	↑	↓	19505
D	8	6	↓	↓	↓	↓	↓	↓	5297	↑	↓	↓	19505
E	13	9	↓	↓	↓	↓	↓	3344	↑	↓	↓	11456	19590
F	20	13	↓	↓	↓	↓	2273	↑	↓	↓	7250	12454	17633
G	32	20	↓	↓	↓	1520	↑	↓	↓	4502	7660	10857	17244
H	50	31	↓	↓	1042	↑	↓	↓	2912	4889	6899	10950	17029[2]
J	80	48	↓	688	↑	↓	↓	1847	3063	4302	6806	10585	15627[1]
K	125	70	457	↑	↓	↓	1196	1969	2754	4342	6745	9961[6]	14790[1]
L	200	98	292	↑	↓	754	1234	1721	2706	4199	6199	9210[2]	↑
M	315	139	↓	485	789	1097	1720	2663	3927	5832	↑	↑	↑
N	500	181	307	498	691	1082	1673	2467	3662	↑	↑	↑	↑
P	800	232	193	312	432	676	1044	1538	2283	↑	↑	↑	↑
Q	1250	286	200	277	432	667	982	1457	↑	↑	↑	↑	↑
R	2000	351	173	270	416	613	909	↑	↑	↑	↑	↑	↑

NOTE: Figures in parentheses indicate the maximum number of independent attributes for which the band of operating characteristic curves is narrow. See section 4.3 for details.

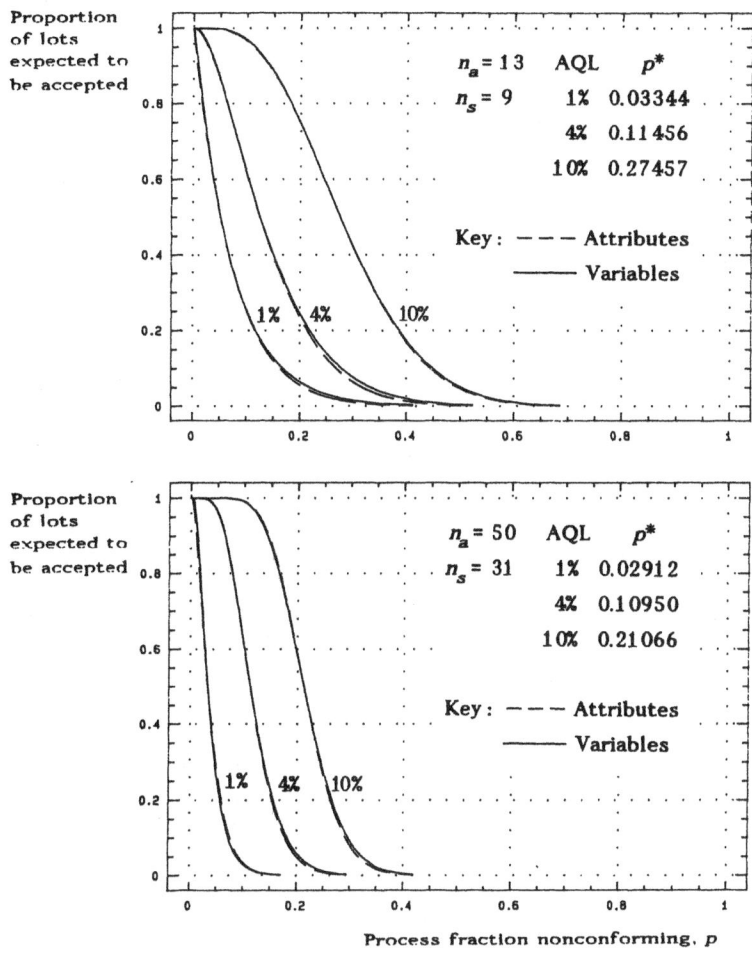

Figure 1. OC curves for single attribute plans and matching single variables plans
("*s*" method with a single specification limit)

follows. The variables sample size n_s corresponding to a whole row of the master table was determined from the attributes sample size n_a by reference to Table 1. With known n_s it was then possible to determine acceptance and rejection constants for each plan independently. The acceptance constant was found by matching the variables OC curve to the OC curve of the attributes plan defined by the sample size n_a and the acceptance number Ac. The rejection constant was found similarly with Ac replaced by $Re - 1$, where Re is the rejection number of the attributes plan. The reduced inspection variables plans matching those of Table II-C of ISO 2859-1:1989 are given in Table 3.

It was also necessary to determine limit constants corresponding to the limit numbers in Table VIII of ISO 2859-1:1989. The limit numbers in this table were calculated as the integral part of the lower 10% points of the normal approximation to the Poisson distribution of the

Table 3. Single Sampling Attriables ("s" Method) Plans for Reduced Inspection (Master Table)

Code letter	n_a	n_s	\multicolumn{11}{c}{Acceptable Quality Levels (Normal Inspection) in % $10^5 p^*$}										
			0.10	0.15	0.25	0.40	0.65	1.0	1.5	2.5	4.0	6.5	10.0
A	2	3										31897	
B	2	3									31897		
C	2	3								31897			31897 / 68103
D	3	3							19524			19524 / 50000	50000 / 80476[1]
E	5	4							8436		8436 / 32362	32362 / 50000[2]	32362 / 67638[1]
F	8	6						5297		5297 / 19505	19505 / 32303	19505 / 44196[1]	32303 / 55804[1]
G	13	9					3344		3344 / 11456	11456 / 19590	11456 / 27457[3]	19590 / 35098[1]	27457[3] / 42587[1]
H	20	13				2273		2273 / 7250	7250 / 12454	7250 / 17633	12454 / 22749[4]	17633 / 27800[1]	27800[1] / 37742[1]
J	32	20			1520		1520 / 4502	4502 / 7660	4502 / 10857	7660 / 14055	10857 / 17244	17244 / 23578[1]	23578[1] / 29856[1]
K	50	31	1042		1042 / 2912	2912 / 4889	2912 / 6899	4889 / 8992	6899 / 10950	10950 / 15005[5]	15005[5] / 19049[1]	21066[1] / 25090[1]
L	80	48		688 / 1847	1847 / 3063	1847 / 4302	3063 / 5551	4302 / 6806	6806 / 9324	9324 / 11846[8]	13107[3] / 15627[1]		
M	125	70	457 / 1196	1196 / 1969	1196 / 2754	1969 / 3546	2754 / 4342	4342 / 5943	5943 / 7548	8352 / 9961[6]			
N	200	98	292 / 754	754 / 1234	754 / 1721	1234 / 2212	1721 / 2706	2706 / 3700	3700 / 4698	5198 / 6199			
P	315	139	485 / 789	485 / 1097	789 / 1408	1097 / 1720	1720 / 2348	2348 / 2978	3294 / 3927				
Q	500	181	307 / 691	498 / 886	691 / 1082	1082 / 1476	1476 / 1871	2070 / 2467					
R	800	232	312 / 554	432 / 676	676 / 921	921 / 1167	1291 / 1538						

NOTE: Figures in parentheses indicate the maximum number of independent attributes for which the band of operating characteristic curves is narrow. See section 4.3 for details.

number of nonconformities. The limit constants were accordingly determined from the formula $(\lambda - 1.2816\sqrt{\lambda})/n_a$ where $\lambda = n_a.AQL\%/100$. The results are given in Table 4. The limit constants are here the limiting acceptable values of the average of $\hat{p}$ over the previous ten lots.

2.3.6 "Attriables" plans

For brevity we shall from now on describe as an attriables plan an acceptance sampling

Table 4. Limit Constants for Reduced Inspection by Attriables

Number of sample units inspected for attributes from last ten lots or batches	Acceptable Quality Level (%)										
	0.10	0.15	0.25	0.40	0.65	1.0	1.5	2.5	4.0	6.5	10.0
	10^5 times limit constant										
20 - 29	*	*	*	*	*	*	*	*	*	*	938
30 - 49	*	*	*	*	*	*	*	*	*	534	2601
50 - 79	*	*	*	*	*	*	*	*	375	1879	4269
80 - 129	*	*	*	*	*	*	*	234	1134	2847	5469
130 - 199	*	*	*	*	*	*	123	723	1752	3634	6445
200 - 319	*	*	*	*	*	94	390	1067	2188	4190	7134
320 - 499	*	*	*	*	72	284	623	1367	2567	4673	7734
500 - 799	*	*	*	38	188	427	798	1594	2854	5039	8188
800 - 1249	*	*	23	113	285	547	945	1784	3094	5345	8567
1250 - 1999	*	10	69	171	358	638	1056	1927	3275	5576	8854
2000 - 3149	9	39	107	219	419	713	1149	2047	3427	5769	9094
3150 - 4999	28	62	136	256	466	772	1220	2139	3543	5918	
5000 - 7999	43	80	159	285	504	819	1278	2213	3638		
8000 - 12499	55	95	178	309	534	857	1325	2273			
12500 - 19999	64	106	193	328	558	885	1360				
20000 - 31499	71	115	205	343	577	909					
31500 - 49999	77	122	214	354	592						
50000 & Over	82	128	221	364	604						

* Denotes that the number of sample units inspected for attributes from the last ten lots or batches is not sufficient for reduced inspection for this AQL. In this instance more than ten lots or batches may be used for the calculation, provided that the lots or batches used are the most recent ones in the sequence, that they have all been on normal inspection, and that none has been rejected while on original inspection.

plan for assessing a class of quality characteristics for which:

(i) at least one of the class is a variable, and at least one an attribute;

(ii) the acceptance criterion is based on the minimum variance unbiased estimator of the overall process fraction nonconforming for the class; and

(iii) the variables sample size n_v and the acceptance constant p^* for the plan have been determined by matching the OC curve of the variables plan (n_v, p^*) to the attributes plan (n_a, Ac) with respect to n_v and p^* subject to (8).

2.4 Partition-Invariance of Attriables Plans under Independence

The probability of accepting an inspection lot on original inspection under independence of the attribute and the variable for given p_a and p_v is

$$P(\hat{p} \le p^* \mid p_a, p_v) = P(1 - \hat{p} \ge 1 - p^* \mid p_a, p_v)$$

$$= P\{(1 - \hat{p}_a)(1 - \hat{p}_s) \ge 1 - p^* \mid p_a, p_v\} \qquad \text{from (4)}$$

$$= P\{1 - \hat{p}_s \ge (1 - p^*)/(1 - \hat{p}_a) \mid p_a, p_v\}$$

$$= P\{\hat{p}_s \le (p^* - \hat{p}_a)/(1 - \hat{p}_a) \mid p_a, p_v\}$$

which, from (5),

$$= P\{\hat{p}_s \le (n_a p^* - r)/(n_a - r) \mid p_a, p_v\}$$

$$= P\{\hat{p}_s \le p_r^* \mid p_a, p_v\}, \qquad (10)$$

say, where $p_r^* = (n_a p^* - r)/(n_a - r)$.

Consider first the case of a single specification limit on the variable. It is shown in the Appendix B that

$$P(\hat{p}_s \le p_r^* \mid p_v) = P\{t \ge (n_s - 1)(1 - 2\beta_{p_r^*}) \mid p_v\}$$

where t is a non-central t-variate with $n_s - 1$ degrees of freedom and eccentricity parameter $\sqrt{n_s}\, K_{p_v}$, K_p being defined by $\Phi(K_p) = 1 - p$ where $\Phi(.)$ is the standard normal distribution function, and β_{p^*} being defined by $B(\beta_{p^*}) = p^*$. Also the statistic r has a binomial distribution with index n_a and probability parameter p_a. From the earlier discussion in 2.3.2 and from (10) it is clear that an inspection lot may not be accepted if r exceeds $[n_a p^*]$, the integral part of $n_a p^*$. Hence

$$P(\hat{p} \le p^* \mid p_a, p_v) = \sum_{r=0}^{[n_a p^*]} \left\{ \binom{n_a}{r} p_a^r (1 - p_a)^{n_a - r} P\{t \ge (n_s - 1)(1 - 2\beta_{p_r^*})\} \right\}. \qquad (11)$$

By varying p_a and p_v subject to (1), upper and lower bounds on the probability of acceptance may be computed for any given process fraction nonconforming p. By varying p from zero up to a suitable fraction, the band of OC curves may be established for any attriables plan under independence. Figure 2 shows the OC bands for the combined plans of Figure 1. It can be seen that the attriables plans have bands of OC curves hardly any broader than the gap between the constituent individual variables and attributes plans. Where the OC curves for these individual plans are well matched, the attriables band of OC curves is consequently quite narrow.

For the case of combined double specification limits on the variable, it is known that the bands of OC curves for variables plans are very narrow, and for most practical purposes may be considered to be unique curves. See, for example, Figure 1 of Baillie (1987). It is therefore to be expected that attriables bands of OC curves for combined double specification limits on the variable would be only slightly broader than the corresponding bands for a single specification limit, under independence.

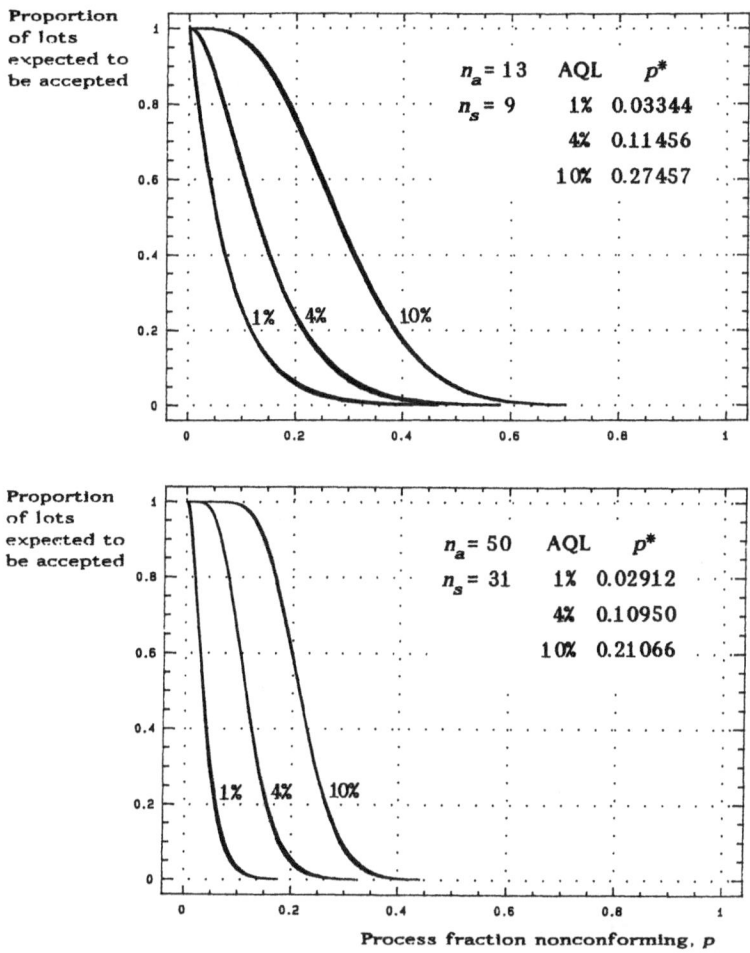

Figure 2. Bands of OC curves for attriables plans with a single attribute and a single independent variable ("s" method with a single specification limit)

We conclude that "s" method attriables plans under independence are highly partition-invariant.

3.   ONE ATTRIBUTE AND ONE DEPENDENT VARIABLE

3.1 The Overall Process Fraction Nonconforming, p, under Dependence

In this section we again consider one variable and one attribute, but this time the variable has a distribution that depends on whether or not the attribute conforms to specification. We assume that when the attribute does conform, the variable has a normal distribution with unknown mean and variance, but that when the attribute does not conform, the variable has some unspecified distribution. This rather vague form of dependence was investigated because it is mathematically tractable and may suggest what to anticipate from more specific forms of dependence.

Denote the overall process fraction nonconforming by p and the process fraction nonconforming for the attribute by p_a, as before. However, this time we shall need the *conditional* process fraction nonconforming for the variable given that the attribute conforms, which we shall denote by $p_{v|a}$. The overall process fraction conforming equals the product of the process fraction conforming for the attribute multiplied by the conditional process fraction conforming for the variable, i.e.

$$1-p = (1-p_a)(1-p_{v|a})$$

which on rearranging gives

$$p = 1 - (1-p_a)(1-p_{v|a}). \qquad (12)$$

3.2 Attriables Sampling Procedure under Dependence

The proposed sampling procedure is a little complicated because:

(a) measurements are only required on attribute-conforming sample units;

(b) precisely n_s measurements are required; and

(c) at least n_a sample units must be classified with respect to the attribute.

Plans with $n_a < n_s$ make little sense in this situation, so it is suggested that, under dependence between the variables and the attributes, sample size code letter A should be subsumed by sample size code letter B for normal and tightened inspection, and A, B and C should be subsumed by D under reduced inspection.

It will be shown in section 3.4 that if r exceeds

$$c = \begin{cases} [n_a p^*] & \text{when } n_s \le n_a - [n_a p^*] \\ [(n_s-1)p^*/(1-p^*)] & \text{otherwise,} \end{cases}$$

then it is impossible to satisfy the acceptance criterion.

An example of the sample space is given in Figure 3a for the case $[n_a p^*] \le n_a - n_s$ and in Figure 3b for the complementary case.

The sampling procedure can be described in words as follows:

Select sample units from a lot at random, classifying them with respect to the attribute into those that are attribute-conforming and those that are attribute-nonconforming until sampling has yielded

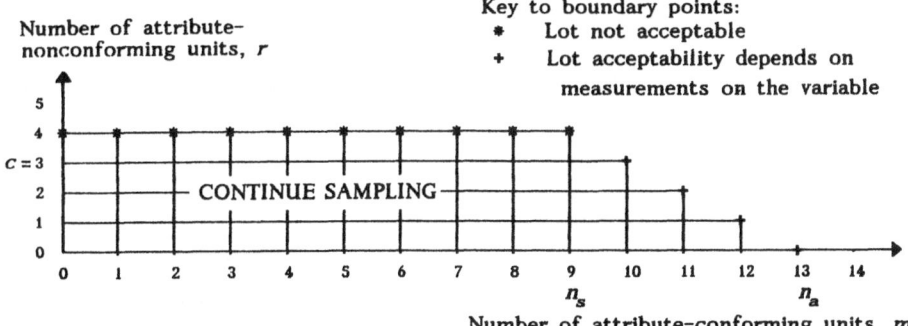

Figure 3a. Example of Sample Space for Dependent "s" Method Attriables Inspection with $[n_a p^*] \le n_a - n_s$: $n_a = 13$, $n_s = 9$, $p^* = 0.27457$, $[n_a p^*] = 3$, $c = 3$.

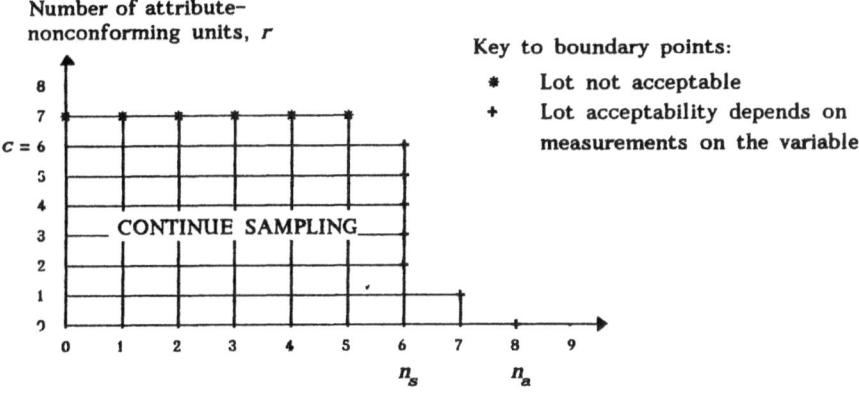

Figure 3b. Example of Sample Space for Dependent "s" Method Attriables Inspection with $[n_a p^*] > n_a - n_s$: $n_a = 8$, $n_s = 6$, $p^* = 0.55804$, $[n_a p^*] = 4$, $c = 6$.

    (i) $c + 1$ attribute-nonconforming units, or

    (ii) n_s attribute-conforming units,

whichever comes first.

In case (i) the lot may be rejected without any calculation or further sampling. (However, for process monitoring purposes it is recommended that a sample of at least n_a units should be drawn, on each of which the attribute should be classified and on each attribute-conforming unit of which the variable should be measured.)

In case (ii) measure the variable on each of the n_s attribute-conforming units sampled so far. Finally, if less than n_a units have been sampled altogether, continue sampling and classifying the attribute until the total sample size reaches n_a.

Lot acceptability in case (ii) is determined by calculating $\hat{p}$ in accordance with section 3.3 and comparing $\hat{p}$ with the applicable value of p^*.

3.3 The MVU Estimator of p under Dependence

Boundary points in Figures 3a and 3b that are on the diagonal $m + r = n_a$ are reached by ordinary binomial sampling, for which it was shown by Girshick *et al* (1946) that

$$\hat{p}_a = r/n_a \qquad (13)$$

is the unique unbiased estimator of p_a based on the sufficient statistic. Boundary points in Figure 3b on the vertical $m = n_s$ for $r > n_a - n_s$ are reached by inverse binomial sampling, for which Girshick *et al* (1946) showed that the unique unbiased estimator of p_a based on the sufficient statistic is, in the present notation,

$$\hat{p}_a = r/(r + n_s - 1). \qquad (14)$$

By similar reasoning to that given in 2.2 for the independent case, the MVU estimator of p is found to be

$$\hat{p} = 1 - (1 - \hat{p}_a)(1 - \hat{p}_{s|a}) \qquad (15)$$

$$= \begin{cases} 1 - (1 - r/n_a)(1 - \hat{p}_{s|a}) & \text{for } r = 0, 1, ..., n_a - n_s \\ 1 - \{1 - r/(r + n_s - 1)\}(1 - \hat{p}_{s|a}) & \text{for } r = n_a - n_s + 1, \ n_a - n_s + 2, ... \end{cases}$$

from (13) and (14) . The quantity $\hat{p}_{s|a}$ is the "s" method MVU estimator of $p_{v|a}$, which is of the form (6) with $\bar{x}$ and S computed for the variable from the measurements taken on the attribute-conforming units.

3.4 Partition-Invariance of "s" Method Attriables Plans under Dependence

For acceptance regions of the form $\hat{p} \leq p^*$, the probability of accepting an inspection lot on original inspection under dependence of the attribute and the variable for given p_a and $p_{v|a}$ is

$$P(\hat{p} \leq p^* | p_a, p_{v|a}) = P(1 - \hat{p} \geq 1 - p^* | p_a, p_{v|a})$$

$$= P\{(1 - \hat{p}_a)(1 - \hat{p}_{s|a}) \geq 1 - p^* | p_a, p_{v|a}\} \quad \text{from (15)}$$

$$= P\{\hat{p}_{s|a} \leq (p^* - \hat{p}_a)/(1 - \hat{p}_a) | p_a, p_{v|a}\}$$

$$= P(\hat{p}_{s|a} \leq p_r^* | p_a, p_{v|a}), \qquad (16)$$

say, where

$$p_r^* = (p^* - \hat{p}_a)/(1 - \hat{p}_a). \tag{17}$$

Using (13) and (14) and rearranging gives

$$p_r^* = \begin{cases} p^* - r(1-p^*)/(n_a - r) & \text{for } r = 0, 1, ..., n_a - n_s \\ p^* - r(1-p^*)/(n_s - 1) & \text{for } r = n_a - n_s + 1, n_a - n_s + 2, ... \end{cases} \tag{18}$$

It is shown in Appendix C that $\{p_r^*\}$ is a strictly monotonic decreasing sequence whose last non-negative element is p_c^* where c is given by

$$c = \begin{cases} [n_a p^*] & \text{when } n_s \leq n_a - [n_a p^*], \\ [(n_s - 1)p^*/(1-p^*)] & \text{otherwise.} \end{cases} \tag{19}$$

Let $c^0 = \min(c, n_a - n_s)$. Then, from (16),

$$P(\hat{p} \leq p^* \mid p_a, p_{v|a}) = \sum_{r=0}^{n_a - n_s} \binom{n_a}{r} p_a^r (1-p_a)^{n_a - r} P(\hat{p}_{s|a} \leq p_r^* \mid p_{v|a})$$

$$+ \sum_{r=n_a - n_s + 1}^{\infty} \binom{r + n_s - 1}{r} (1-p_a)^{n_s} p_a^r P(\hat{p}_{s|a} \leq p_r^* \mid p_{v|a}) \tag{20}$$

$$= \sum_{r=0}^{c^0} \binom{n_a}{r} p_a^r (1-p_a)^{n_a - r} P\{t \geq (n_s - 1)(1 - 2\beta_{p_r^*})\}$$

$$+ \sum_{r=n_a - n_s + 1}^{c} \binom{r + n_s - 1}{r} (1-p_a)^{n_s} p_a^r P\{t \geq (n_s - 1)(1 - 2\beta_{p_r^*})\} \tag{21}$$

where t is the noncentral t-variate with $n_s - 1$ degrees of freedom and eccentricity parameter $\sqrt{n_s} \, K_{p_{v|a}}$

By varying p_a and $p_{v|a}$ subject to (12), the range of values of the probability of accepting a lot for a given value of p^* can be computed for a given value of p, the process fraction nonconforming. Repeating this for a range of values of p from zero up to a suitable fraction yields the OC band under dependence.

Happily it turns out that there is no difference between the independent and dependent variants of the OC band for most practical combinations of n_a, n_s and p^*. For compare the OC band generated from (20) by varying p_a and $p_{v|a}$ subject to (12) with the OC band generated from (11) by varying p_a and p_v subject to (1). The OC bands will be identical provided

$$n_a - n_s > [n_a p^*] \tag{22}$$

or

$$n_a - n_s = [n_a p^*] \quad \text{and} \quad p_{n_a - n_s + 1}^* < 0. \tag{23}$$

18

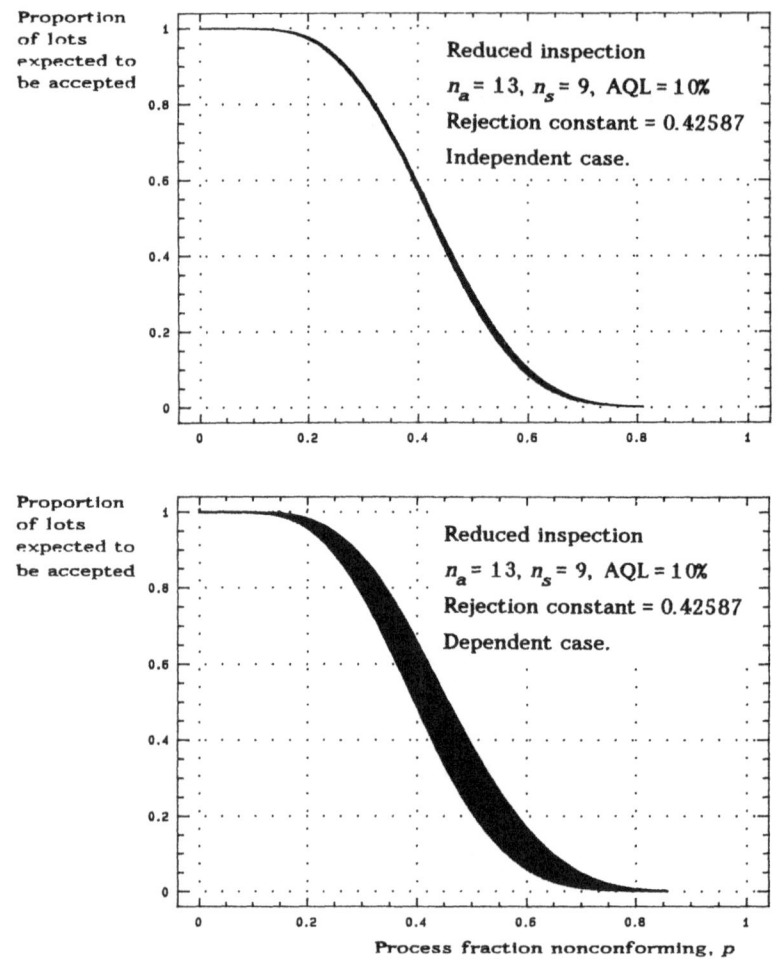

Figure 4. Bands of attriables OC curves for a single attribute and a single variable (*"s"* method with a single specification limit) under independence and under dependence when $n_a - n_s < [n_a p^*]$

Using (18), the inequality in (23) is found to reduce to $n_a - n_s > n_a p^* - 1$, which must hold true when $n_a - n_s = [n_a p^*]$. The inequality part of (23) is therefore redundant. It follows from (22) and (23) that the OC bands under independence and under dependence will be identical whenever

$$n_a - n_s \geq [n_a p^*]. \tag{24}$$

This will be true in general except in the case of small attributes sample sizes in combination with large acceptable quality levels (AQLs). Table 5 shows that none of the six cases depicted in Figures 1 and 2 has an OC band that differs between independence and dependence. Figure 4 shows that the attriables OC band for the reduced inspection plan $(n_a, n_s, p^*) = (13, 9, 0.42587)$ is considerably broader under dependence than under independence.

Table 5. Values of $n_a - n_s - [n_a p^*]$ for the Six Cases of Figures 1 and 2

Case	n_a	AQL	n_s	p^*	$[n_a p^*]$	$n_a - n_s - [n_a p^*]$
1	13	1.0%	9	0.03344	0	4
2	13	4.0%	9	0.11456	1	3
3	13	10.0%	9	0.27457	3	1
4	50	1.0%	31	0.02912	1	18
5	50	4.0%	31	0.10950	5	14
6	50	10.0%	31	0.21066	10	9

4. MULTIVARIATE GENERALIZATIONS

4.1 Variables Independent of the Attributes

Suppose that there are at least three quality characteristics of which at least one is an attribute and at least one a variable, and for which the variables are independent of the attributes. Suppose also that neither the means nor the variances of the variables are known.

The sampling procedure is a simple generalization of that given in Section 2. A random sample of n_s items is drawn for the purpose of inspecting the variables *and* the attributes, and a further random sample of size $n_a - n_s$ is drawn for the purpose of inspecting the attributes alone.

Let p_v and p_a now more generally denote the combined process fraction nonconforming for the variables and for the attributes respectively. An elementary generalization of the lemma in Appendix A shows that expression (4) remains true, i.e. $\hat{p} = 1 - (1 - \hat{p}_s)(1 - \hat{p}_a)$ with $\hat{p}_s$ and $\hat{p}_a$ now defined respectively as the MVU estimators of the process fractions nonconforming for the subclasses containing the variables and the attributes respectively.

Suppose further that there are m variables and M attributes. If the m variables are mutually independent and the M attributes are mutually independent, then

$$\hat{p} = 1 - \{ \prod_{i=1}^{m} (1 - \hat{p}_{s_i}) \} \{ \prod_{j=1}^{M} (1 - \hat{p}_{a_j}) \} \tag{25}$$

where $\hat{p}_{s_i}$ and $\hat{p}_{a_j}$ are the univariate MVU estimators of the process fraction nonconforming for the ith variable and the jth attribute respectively, being of the form (6) and (5).

If, on the other hand, the m variables are known to form q mutually independent groups and the M attributes to form Q mutually independent groups, then

$$\hat{p} = 1 - \left\{ \prod_{j=1}^{q} (1 - \hat{p}'_{s_j}) \right\} \left\{ \prod_{j=1}^{Q} (1 - \hat{p}'_{a_j}) \right\} \tag{26}$$

where $\hat{p}'_{s_j}$ and $\hat{p}'_{a_j}$ represent the MVU estimators of the process fraction nonconforming for the ith group of variables and for the jth group of attributes respectively. The estimator $\hat{p}'_a$

is evaluated by counting the number of items in the sample of size n_a that have at least one nonconformity among the jth group of attributes, and dividing by n_a. The formula for multivariate MVU estimators by variables for unknown process means and unknown process covariance matrix (i.e. the multivariate "s" method) is given by Baillie (1987). For example, if $q = 1$ so that the m variables can not be separated into two or more mutually independent groups, then

$$\hat{p}_s = \hat{p}_{s_1} = 1 - \frac{\Gamma\{\frac{1}{2}(n_s - 1)\}\,|\mathbf{R}|^{-\frac{1}{2}}}{\Gamma\{\frac{1}{2}(n_s - 1 - m)\}\,\pi^{m/2}} \int\limits_{Z_m}\cdots\int \left(1 - \mathbf{z}'\mathbf{R}^{-1}\mathbf{z}\right)^{\frac{1}{2}(n_s - m - 3)} \prod_{j=1}^{m} dz_j \tag{27}$$

where $\mathbf{R}$ is the sample product-moment correlation matrix and Z_m is the region of intersection of the hyperellipsoid $\mathbf{z}'\mathbf{R}^{-1}\mathbf{z} \le 1$ with the m-fold rectangle $\mathbf{l} \le \mathbf{z} \le \mathbf{u}$ where

$$l_j = (L_j - \overline{x}_j)/\left\{S_j\,\sqrt{(n_s - 1)/n_s}\right\} \quad \text{and} \quad u_j = (U_j - \overline{x}_j)/\left\{S_j\,\sqrt{(n_s - 1)/n_s}\right\}$$

for $j = 1, 2, ..., m$, where L_j, U_j are the lower and upper specification limits and $\overline{x}_j$, S_j^2 the sample mean and the sample sum of squares about the mean for the jth variable.

4.2 Some Variables Dependent on Attributes

Suppose that the attributes and variables may each be divided into k groups such that
(a) the first group of attributes is unrelated to attributes or variables in any other groups;
(b) the first group of variables is unrelated to attributes or variables in any other groups;
(c) every variable in the ith group of variables is related to one or more attributes in the ith group of attributes, but is not related to attributes in other groups, for $i = 2$, 3, ..., k; and
(d) every attribute in the ith group of attributes is related to one or more variables in the ith group of variables, but is not related to variables in other groups, for $i = 2, 3, ..., k$.
In the above context, two quality characteristics are "related", and must therefore appear in like-numbered groups, if nonconformity of one precludes observation of the other.

The first group of attributes or variables may be a null group. In practice, cases with values of k much in excess of 2 will be rare. Denote the ith group of attributes by A_i and the ith group of variables by V_i, with corresponding process fractions nonconforming p_{A_i} and p_{V_i}, for $i = 1, 2, ..., k$. Then an obvious generalization of (15) is

$$\hat{p} = 1 - (1 - \hat{p}_{A_1})(1 - \hat{p}_{V_1}) \prod_{i=2}^{k} \left\{(1 - \hat{p}_{A_i})(1 - \hat{p}_{V_i|A_i})\right\}.$$

$$= 1 - \left\{\prod_{i=1}^{k}(1 - \hat{p}_{A_i})\right\}(1 - \hat{p}_{V_1}) \prod_{j=2}^{k}(1 - \hat{p}_{V_j|A_j}). \tag{28}$$

If the sampling procedures of 2.3.1 and 3.2 are appropriately combined, then it is reasonable to expect that plans with acceptance criteria of the form $\hat{p} \leq p^*$ will be fairly partition-invariant. The appropriate acceptance inspection procedure is as follows:

(i) Select a sample unit at random from the lot or batch;

(ii) Classify all the attributes; measure the variables in group 1 provided no more than n_s units have already been measured in this group;

(iii) Measure all the variables in the ith group of variables provided there are no nonconformities in the ith group of attributes and no more than n_s units have already been measured in this group, for $i = 2, 3, ..., k$;

(iv) Repeat from (i) until sampling has yielded n_s measured units for each group of variables, or a total of $c + 1$ attribute-nonconforming sample units, whichever comes first. In the former case, if less than n_a sample units have been selected altogether, continue sampling units and classifying the attributes until the total sample size reaches n_a; then calculate $\hat{p}$ from (28) and accept the lot if and only if $\hat{p} \leq p^*$. In the latter case the lot may be rejected without any calculation or further sampling, although further sampling may be deemed advisable for process monitoring purposes.

Example

Consider sampling inspection of 8 quality characteristics, four of which are attributes and four are variables. There are three groups of attributes and variables. The first, independent, group consists of one attribute and one variable; the second group consists of two attributes and one variable; and the third group consists of one attribute and two variables. Thus, symbolically, the product-moment correlation matrix between the eight quality characteristics is known to be of the form:

$$
\begin{array}{c}
\begin{array}{c} a_1 \\ v_1 \\ \\ a_2 \\ a_3 \\ v_2 \\ \\ a_4 \\ v_3 \\ v_4 \end{array}
\left[
\begin{array}{ccc}
\overset{a_1 \ v_1}{\mathbf{I}} & \overset{a_2 \ a_3 \ v_2}{\mathbf{0}} & \overset{a_4 \ v_3 \ v_4}{\mathbf{0}} \\
\\
\mathbf{0} & \Omega_2 & \mathbf{0} \\
\\
\mathbf{0} & \mathbf{0} & \Omega_3
\end{array}
\right]
\end{array}
$$

where $\mathbf{I}$ represents an identity matrix, $\mathbf{0}$ a null matrix and Ω_i the unknown product-moment correlation matrix for the ith group of attributes and variables, $i = 2, 3$. Normal inspection is to be used, with sample size code letter E and an AQL of 10%.

From Table 1 it is found that the attributes sample size for code letter E under normal inspection is $n_a = 13$ and that the corresponding variables sample size and acceptance constant for an AQL of 10% are n_s = 9 and p^* = 0.27457. The acceptance number for a single attribute plan is $[n_a p^*]$, i.e. 3.   As $n_s < n_a - [n_a p^*]$ it follows from (19) that $c = [n_a p^*] = 3$.

A hypothetical sampling sequence is shown below, with √ and × representing conformity and nonconformity of an attribute and with $x_{i,j}$ representing the jth measurement taken on variable i. The symbol ▬ indicates that the variable has not been measured, either because n_s measurements have already been obtained on this variable or because there are one or more nonconformities on this sample unit among the corresponding group of attributes. Note that as a result the jth measurement on variable i is not necessarily taken on the jth sample unit.

		Unit Sampled												
Group	Attribute	1	2	3	4	5	6	7	8	9	10	11	12	13
1	1	√	×	√	√	√	√	√	√	√	√	√	√	√
2	2	√	√	√	√	√	√	√	√	√	√	√	√	√
	3	√	√	√	×	√	√	√	√	√	√	√	√	√
3	4	√	√	√	√	√	√	×	√	√	√	√	√	√

Group	Variable	1	2	3	4	5	6	7	8	9	10	11	12	13
1	1	$x_{1,1}$	$x_{1,2}$	$x_{1,3}$	$x_{1,4}$	$x_{1,5}$	$x_{1,6}$	$x_{1,7}$	$x_{1,8}$	$x_{1,9}$	▬	▬	▬	▬
2	2	$x_{2,1}$	$x_{2,2}$	$x_{2,3}$	▬	$x_{2,4}$	$x_{2,5}$	$x_{2,6}$	$x_{2,7}$	$x_{2,8}$	$x_{2,9}$	▬	▬	▬
3	3	$x_{3,1}$	$x_{3,2}$	$x_{3,3}$	$x_{3,4}$	$x_{3,5}$	$x_{3,6}$	▬	$x_{3,7}$	$x_{3,8}$	$x_{3,9}$	▬	▬	▬
	4	$x_{4,1}$	$x_{4,2}$	$x_{4,3}$	$x_{4,4}$	$x_{4,5}$	$x_{4,6}$	▬	$x_{4,7}$	$x_{4,8}$	$x_{4,9}$	▬	▬	▬

It may be possible to effect inspection cost savings by not measuring certain groups of variables until sampling has been completed. This will be the case when not all of the measurements on sample units have to be made at the same time as the classification of the attributes on those units, as the attributes results alone may be sufficient to determine that the lot is not acceptable. This will be particularly true when the unit costs of the measurements are high and process quality is at or worse than the AQL.

4.3 Partition-Invariance of "s" Method Attriables Plans

Baillie (1987) showed that the multivariate independent "s" method is approximately partition-invariant, and conjectured that similar results obtain for the multivariate dependent "s" method. Baillie (1991) showed that k-attribute acceptance sampling plans under independence are approximately partition-invariant if

$$c/n_a \leq p^* < 1 - \{1 - c_k/n_a\}^{k-c-1+kc_k} \{1-(1+c_k)/n_a\}^{c+1-kc_k} \qquad (29)$$

where c is the acceptance number of the corresponding single attribute plan and c_k is the integral part of $(c+1)/k$. Most of the plans in Tables 1, 2 and 3 (and the corresponding Tables 6, 7 and 8 for the "σ" method) satisfy (29) for any number of attributes, i.e. for any positive integral value of k; those that do not have been annotated to show the maximum number of attributes for which (29) is true. There is of course a unique OC curve for dependent attribute fraction nonconforming plans. All these results, taken in conjunction with the high degree of partition-invariance of attriables plans consisting of one attribute and one variable regardless of dependence, demonstrated above, suggest that attriables plans consisting of several attributes in combination with several normally distributed variables with unknown means and covariance matrix will be reasonably partition-invariant.

Where necessary, the acceptance constant could have been constrained further in the matching procedure in order to satisfy (29) over a wider range of values of k, but this was not done as it would have compromised the partition-invariance for the lower values of k.

5. ATTRIABLES PLANS WHEN THE COVARIANCE MATRIX OF THE VARIABLES IS KNOWN

5.1 Matching Variables and Attributes Plans when the Covariance Matrix of the Process Variables is Known

For univariate sampling by variables it is well known (e.g. Lieberman and Resnikoff, 1955) that the sample size n_σ of known process standard deviation (i.e. "σ" method) plans is less than or equal to the sample size n_s of matching unknown process standard deviation (i.e. "s" method) plans. In other words, knowledge of the process standard deviation in general permits similar levels of discrimination between good and bad quality to be achieved with a lower sample size. In the case of attriables plans when the process covariance matrix of the variables is known, the matching, as before, must be carried out subject to a common value of p^*. The matching to the attributes plans was carried out in a similar way to that described for the "s" method in 2.3. Tables 6, 7 and 8 show the resulting "σ" method variables plans which respectively match the normal, tightened and reduced inspection plans of ISO 2859-1: Part 1, with Table 4 giving the limit constants for $\hat{p}_\sigma$ which correspond to the limit numbers shown in Table VIII of the ISO standard. Figure 5 shows how good the matching is for a selection of plans from Table 6, with a single specification limit on the variable.

Table 6. Single Sampling Attriables ("σ" Method) Plans for Normal Inspection (Master Table)

Sample size code letter	Attributes sample size, n_a	Variables sample size, n_σ	Acceptable Quality Levels (Normal Inspection) in % $10^5 p^*$										
			0.10	0.15	0.25	0.40	0.65	1.0	1.5	2.5	4.0	6.5	10.0
A	2	2										22045	
B	3	2									12329		
C	5	3								8337			27622
D	8	4							5486			16673	29516
E	13	6						3740			10457	17874	25659
F	20	8					2558			6887	11570	16445	26537[5]
G	32	11				1684			4399	7285	10262	16384	22652[5]
H	50	20			1188			2995	4873	6786	10674	14617	20598[2]
J	80	27		762			1902	3078	4270	6687	9131	12830	17806[2]
K	125	37	500			1238	1994	2760	4306	5865	8221	11384[9]	16957[1]
L	200	44			777	1249	1727	2690	3661	5126	7091	10553[4]	
M	315	51		495	795	1098	1708	2322	3249	4491	6678		
N	500	59	313	502	692	1076	1463	2045	2825	4200			
P	800	75	316	436	677	918	1283	1770	2627				
Q	1250	84	279	433	588	821	1132	1680					
R	2000	93	271	367	513	707	1050						

Table 7. Single Sampling Attriables ("σ" Method) Plans for Tightened Inspection (Master Table)

Sample size code letter	Attributes sample size, n_a	Variables sample size, n_s	Acceptable Quality Levels (Normal Inspection) in % $10^5 p^*$										
			0.10	0.15	0.25	0.40	0.65	1.0	1.5	2.5	4.0	6.5	10.0
A	2	2											
B	3	2										12329	
C	5	3									8337		
D	8	4								5486			16673
E	13	6							3740			10457	17874
F	20	8						2558			6887	11570	16445
G	32	11					1684			4399	7285	10262	16384
H	50	20				1188			2995	4873	6786	10674	16603
J	80	27			762			1902	3078	4270	6687	10360	15313[3]
K	125	37		500			1238	1994	2760	4306	6649	9800	14564[2]
L	200	44	315			777	1249	1727	2690	4148	6107	9066	
M	315	51			495	795	1098	1708	2630	3869	5739		
N	500	59		313	502	692	1076	1656	2434	3608			
P	800	75	197	316	436	677	1040	1526	2260				
Q	1250	84	203	279	433	665	976	1445					
R	2000	93	175	271	416	610	902						

NOTE: Figures in parentheses indicate the maximum number of independent attributes for which the band of operating characteristic curves is narrow. See section 4.3 for details.

Table 8. Single Sampling Attriables ("σ" Method) Plans for Reduced Inspection (Master Table)

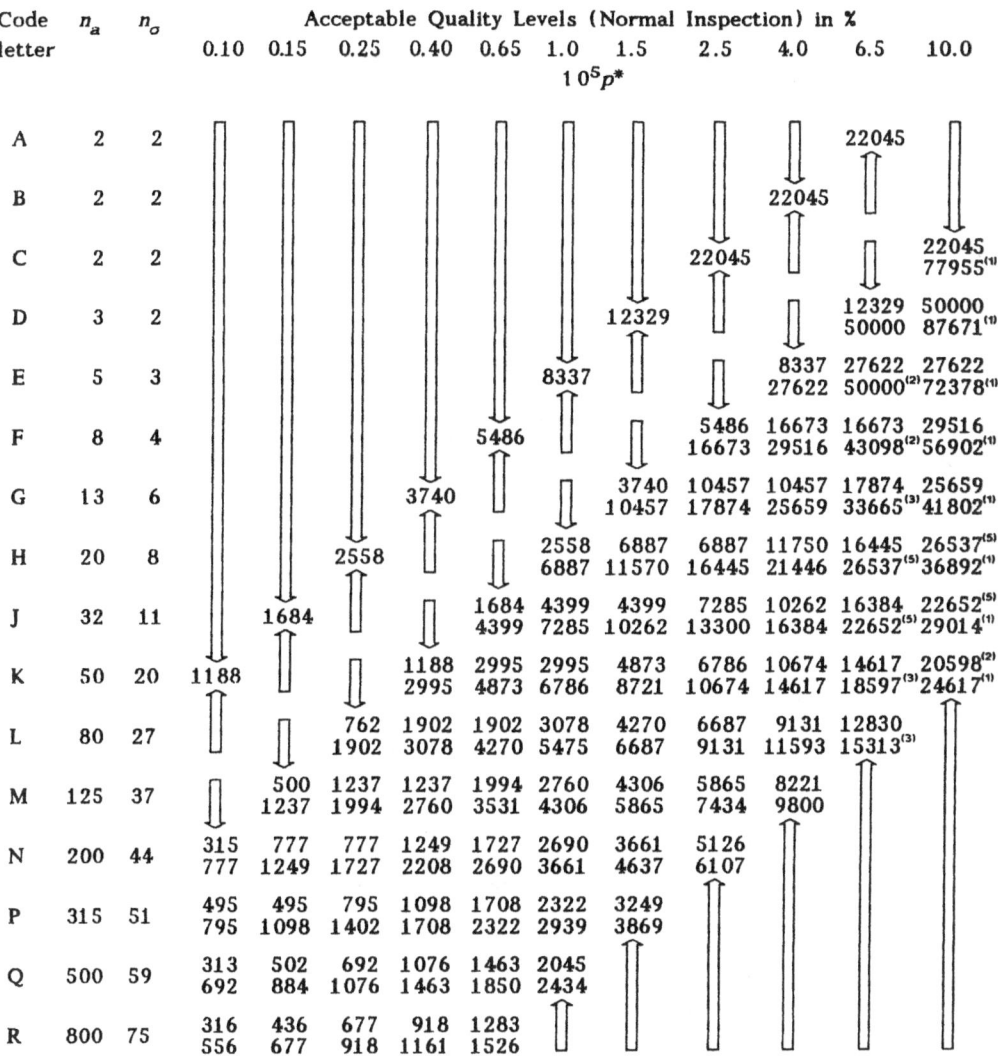

Code letter	n_a	n_σ	0.10	0.15	0.25	0.40	0.65	1.0	1.5	2.5	4.0	6.5	10.0
								$10^5 p^*$					
A	2	2										22045	
B	2	2									22045		
C	2	2								22045			22045 / 77955[1]
D	3	2							12329			12329 / 50000	50000 / 87671[1]
E	5	3						8337			8337 / 27622	27622 / 50000[2]	27622 / 72378[1]
F	8	4					5486			5486 / 16673	16673 / 29516	16673 / 43098[2]	29516 / 56902[1]
G	13	6				3740			3740 / 10457	10457 / 17874	10457 / 25659	17874 / 33665[3]	25659 / 41802[1]
H	20	8			2558			2558 / 6887	6887 / 11570	6887 / 16445	11750 / 21446	16445 / 26537[5]	26537 / 36892[1]
J	32	11		1684			1684 / 4399	4399 / 7285	4399 / 10262	7285 / 13300	10262 / 16384	16384 / 22652[5]	22652 / 29014[1]
K	50	20	1188		1188 / 2995	2995 / 4873	2995 / 6786	4873 / 8721	6786 / 10674	10674 / 14617	14617 / 18597[3]	20598[2] / 24617[1]	
L	80	27			762 / 1902	1902 / 3078	1902 / 4270	3078 / 5475	4270 / 6687	6687 / 9131	9131 / 11593	12830 / 15313[3]	
M	125	37		500 / 1237	1237 / 1994	1237 / 2760	1994 / 3531	2760 / 4306	4306 / 5865	5865 / 7434	8221 / 9800		
N	200	44	315 / 777	777 / 1249	777 / 1727	1249 / 2208	1727 / 2690	2690 / 3661	3661 / 4637	5126 / 6107			
P	315	51	495 / 795	495 / 1098	795 / 1402	1098 / 1708	1708 / 2322	2322 / 2939	3249 / 3869				
Q	500	59	313 / 692	502 / 884	692 / 1076	1076 / 1463	1463 / 1850	2045 / 2434					
R	800	75	316 / 556	436 / 677	677 / 918	918 / 1161	1283 / 1526						

Acceptable Quality Levels (Normal Inspection) in %

NOTE: Figures in parentheses indicate the maximum number of independent attributes for which the band of operating characteristic curves is narrow. See section 4.3 for details.

5.2 Attriables Procedures when the Covariance Matrix of the Process Variables is Known

The formula for $\hat{p}_\sigma$ in the univariate case is given by Lieberman and Resnikoff (1955) as

$$\hat{p}_\sigma = \Phi\left\{\sqrt{n_\sigma(n_\sigma-1)}\,(\bar{x}-U)/\sigma\right\} + \Phi\left\{\sqrt{n_\sigma(n_\sigma-1)}\,(L-\bar{x})/\sigma\right\} \tag{30}$$

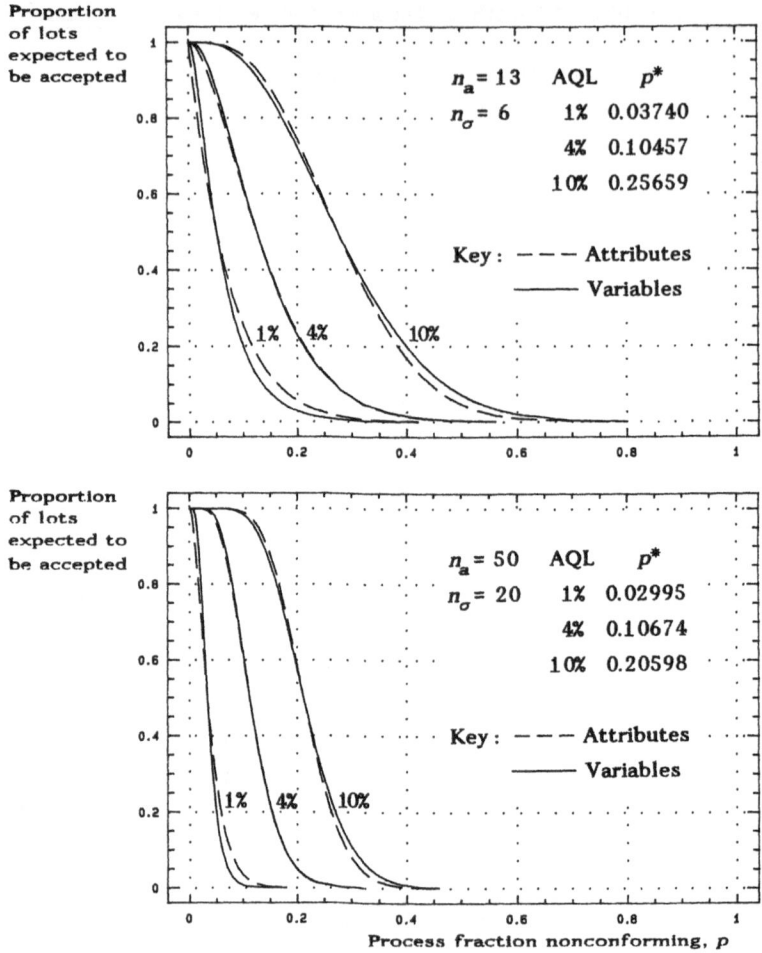

Figure 5. OC curves for single attribute plans and matching single variables plans
("σ" method with a single specification limit)

where σ denotes the process standard deviation, n_σ denotes the sample size and $\Phi(.)$ denotes the standard normal distribution function.

The formula for $\hat{p}_\sigma$ in the multivariate case with m variables is given by Baillie (1987) as

$$\hat{p}_\sigma = 1 - (2\pi)^{-m/2} |\mathbf{P}|^{-\frac{1}{2}} \int_{a_m}^{b_m} \int_{a_{m-1}}^{b_{m-1}} \ldots\ldots\ldots \int_{a_1}^{b_1} \exp\left(-\tfrac{1}{2}\, \mathbf{w}'\, \mathbf{P}^{-1} \mathbf{w}\right) \prod_{j=1}^{m} dw_j \tag{31}$$

where

$\mathbf{P}$ is the process correlation matrix;

$$a_j = (L_j - \overline{x}_j) / \left\{ \sigma_j \sqrt{(n_\sigma - 1)/n_\sigma} \right\}, \quad j = 1, 2, ..., m;$$

and

$$b_j = (U_j - \overline{x}_j) / \left\{ \sigma_j \sqrt{(n_\sigma - 1)/n_\sigma} \right\}, \quad j = 1, 2, ..., m.$$

As the sample mean vector $\overline{x}$ is a complete sufficient statistic for a multivariate normal family of distributions with known covariance matrix (e.g. see Baillie, 1987), Lemma 1 can be easily modified to correspond to the σ-known case. In effect, this means replacing n_s by n_σ, $\hat{p}_s$ by $\hat{p}_\sigma$, (6) by (30) and (27) by (31) in earlier sections to give the corresponding attriables procedures for the case where the covariance matrix of the process variables is known.

Figure 6 shows the OC bands under independence of attriables plans consisting of a "σ" method variable and an attribute, which correspond to the pairs of OC curves given in Figure 5. These bands are broader than those of Figure 2, indicating that "σ" method attriables plans for one attribute and one independent variable are less partition-invariant than their "s" method attriables counterparts.

By analogy with (24) it may be deduced that "σ" method attriables plans under dependence have the same OC bands as under independence, provided

$$n_a - n_\sigma \geq [n_a p^*] \tag{32}$$

where p^* is now a "σ" method acceptance constant. As n_σ is typically smaller than n_s and the acceptance constant p^* is typically of similar magnitude for corresponding "σ" method and "s" method cases, it follows that an even greater range of "σ" method attriables plans satisfy (32) than "s" method attriables plans satisfy (24). In fact, it can be seen that all the "σ" method attriables plans in Tables 6 and 7 and most of those in Table 8 satisfy (32). This is somewhat fortunate in view of the way in which dependence was seen from Figure 4 to inflate the "s" method attriables OC band when (24) was not satisfied.

Figures 4 and 5 and sections 4.3 and 4.4 of Baillie (1987) suggest that the OC band width for multivariate plans increases rapidly with the number of "σ" method variables, so caution should be exercised in interpreting the data from "σ" method attriables plans with more than two or three variables.

6. SUMMARY AND CONCLUSIONS

Acceptance sampling plans have been developed for application to a group or class of attributes and variables that are of about equal criticality to product (or process, or service) acceptability. The plans match those of ISO 2859, the international standard on sampling by attributes. Relationships that are known to exist between the quality characteristics affect the way in which the chosen plan is implemented. In general, the sample size for measuring variables is smaller than the sample size for classifying attributes, which in many cases is advantageous in terms of cost and time. The chief advantage of these plans over existing plans,

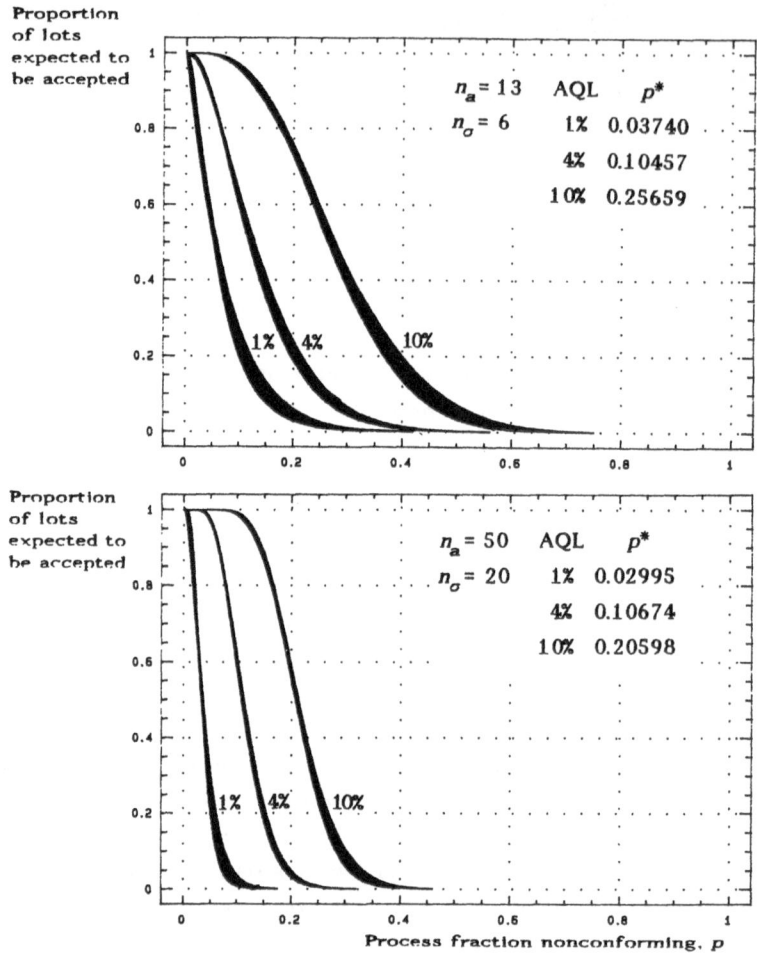

Figure 6. Bands of OC curves for attriables plans with a single attribute and a single independent variable ("σ" method with a single specification limit)

however, is that the probability of acceptance of a lot on original inspection with respect to this class of quality characteristics is more or less the same regardless of how the class-specific process fraction nonconforming is partitioned between the elements of the class, i.e. the plans are reasonably "partition-invariant". This is achieved without resort to the wasteful procedure of reclassifying variables as attributes, and without recourse to AQL apportionment among the characteristics of the class, with its attendant risks. Attriables plans involving variables with unknown variances and covariances turn out to be particularly partition-invariant. Another useful property of the plans is that in the majority of cases the dependent and independent variants have identical bands of operating characteristic curves. Thus little if any advantage is to be had by the consumer or the producer in falsely claiming dependence or independence.

Although this result has only been shown to be true for the particular form of dependence considered, it augurs well for other forms of dependence.

In deriving the plans, a novel feature in matching the variables plans to the attributes plans was the use of a common acceptance constant in order to harmonize the variables and attributes acceptance sampling methodologies. In order to simplify the implementation of attriables plans, the variables sample sizes were constrained to be constant across rows of the master tables, i.e. to be the same regardless of AQL. The matching was focussed on the upper part of the OC curves, which in the majority of cases resulted in smaller variables sample sizes than would have been the case if attention had been given to the whole curve. The penalty for achieving relatively good matching over the top part of the OC curves turned out to be a surprisingly modest lack of fit over the bottom part. As a production process would soon be put onto tightened inspection and then soon have inspection discontinued if quality deteriorated to such levels, this slight penalty was considered inconsequential.

Plans have been developed for attributes in combination with normally distributed variables having unknown means and covariance matrix or in combination with normally distributed variables having unknown means and known covariance matrix. As the acceptance constants differ between the two cases, the plans are not suited to attriables inspection where both types of variable are involved. To accommodate this situation it would be possible to simultaneously fit both types of variables plan to the corresponding attributes plan. Owing to the fundamentally different shapes of "s" method and "σ" method OC curves for the smaller sample sizes, it is anticipated that the resulting three-way attriables plans for the earlier sample size code letters would be substantially less partition-invariant than their two-way counterparts.

The dependent plans are rather too complicated for unsupported use on the shop floor, so it is envisaged that they would be used interactively in conjunction with a specially developed microcomputer software package.

ACKNOWLEDGEMENT

Any views expressed in the paper are those of the author and do not necessarily represent those of H. M. Government.

REFERENCES

[1] BAILLIE, D. H. (1987): Multivariate acceptance sampling,
in H.-J. Lenz, G. B. Wetherill and P.-Th. Wilrich (Eds.) Frontiers in Statistical Quality Control 3,
Physica-Verlag, Heidelberg, Germany.

[2] BAILLIE, D. H. (1991): Multi-attribute acceptance sampling plans under independence.
Proceedings of the 3rd Conference of the Asia Pacific Quality Control Organisation,

Volume 5.

[3] BLACKWELL, D. (1947): Conditional expectation and unbiased sequential estimation, Annals of Mathematical Statistics, 18, 105 - 110.

[4] BOWKER, A. H. and GOODE, H. P. (1952): Sampling Inspection by Variables, McGrawHill, New York.

[5] DANZIGER, L. and PAPP, Z. (1988): Multiple criteria sampling plans for total fraction nonconformance, Journal of Quality Technology, 20, 181-187.

[6] GIRSHICK, M.A., MOSTELLER, F. and SAVAGE, L.J.(1946): Unbiased estimates for certain binomial sampling problems with applications, Annals of Mathematical Statistics, 17, 13-23.

[7] GREGORY, G. and RESNIKOFF, G.J. (1955): Some notes on mixed variables and attributes sampling plans, Technical Report No. 10, Applied Mathematics and Statistics Laboratory, Stanford University, Stanford, California.

[8] ISO 2859-1:1989: Part 1 Sampling procedures for inspection by attributes: Part 1: Sampling plans indexed by acceptable quality level (AQL) for lot-by-lot inspection, International Organization for Standardization, Geneva, Switzerland.

[9] ISO 3951:1989, Sampling procedures and charts for inspection by variables for percent Nonconforming, International Organization for Standardization, Geneva, Switzerland.

[10] LEHMANN, E. L. and SCHEFFÉ, H. (1950): Completeness, similar regions and unbiased estimation, Part 1, Sankhyā, 10, 305-340.

[11] LIEBERMAN, G. J. (1953), Contributions to Sampling Inspection, Ph. D. thesis, Stanford University, Stanford, California.

[12] LIEBERMAN, G. J. and RESNIKOFF, G. J. (1955): Sampling plans for inspection by variables, Journal of the American Statistical Association, 50, 457-516.

APPENDIX A: THE MVU ESTIMATOR OF THE OVERALL FRACTION NONCONFORMING UNDER INDEPENDENCE OF THE ATTRIBUTE AND THE VARIABLE

Lemma   For a normally distributed random variable with unknown mean, μ, and unknown variance, σ^2, and an independently distributed random attribute with unknown probability parameter p_a, unique uniformly MVU estimators $\hat{p}_s$, $\hat{p}_a$ and $\hat{p}$ of the process fraction nonconforming exist for the variable, for the attribute and for overall, and are related by

$$\hat{p} = 1 - (1-\hat{p}_s)(1-\hat{p}_a). \tag{A.1}$$

Proof Let

$$\tilde{p}_v = \begin{cases} 0 \text{ if the variable on the first sample unit conforms to specification} \\ 1 \text{ otherwise,} \end{cases}$$

$$\tilde{p}_a = \begin{cases} 0 \text{ if the attribute on the first sample unit conforms to specification} \\ 1 \text{ otherwise} \end{cases}$$

and

$$\tilde{p} = 1 - (1-\tilde{p}_v)(1-\tilde{p}_a). \tag{A.2}$$

Clearly the unconditional expectations of $\tilde{p}_v$ and $\tilde{p}_a$ are p_v and p_a respectively. It follows from the independence of the variable and the attribute that $\tilde{p}$ is an unbiased estimator of p, for

$$\begin{aligned} E(\tilde{p}) &= 1 - \{1-E(\tilde{p}_v)\}\{1-E(\tilde{p}_a)\} \\ &= 1 - (1-p_v)(1-p_a) \\ &= p \text{ from (1)}. \end{aligned}$$

Thus p_v, $\dot{p}_a$ and p are estimable functions. Now let t denote the sufficient statistics for μ and σ^2, and let r denote the sufficient statistic for p_a. Then provided t and r are complete, it follows from theorems of Blackwell (1947) and Lehmann and Scheffé (1950) that the functions

$$\hat{p} = E(\tilde{p}|t, r), \tag{A.3}$$

$$\hat{p}_s = E(\tilde{p}_v|t) \tag{A.4}$$

and

$$\hat{p}_a = E(\tilde{p}_a|r) \tag{A.5}$$

exist and are the unique uniformly MVU estimators of p, p_v and p_a respectively. But completeness of the sufficient statistics *has* been established by Lehmann and Scheffé (1950) for the binomial family of distributions with fixed sample size, and by Lieberman (1953) for the multivariate normal distribution with unknown mean vector and unknown covariance matrix. Hence

$$\begin{aligned} \hat{p} &= E(\tilde{p}|t, r) \text{ from (A.3)} \\ &= 1 - E\{(1-\tilde{p}_v)(1-\tilde{p}_a)|t, r\} \text{ from (A.2)} \\ &= 1 - \{1-E(\tilde{p}_v|t)\}\{1-E(\tilde{p}_a|r)\} \text{ from the independence} \\ &\qquad\qquad\qquad\qquad\qquad \text{of the variable and the attribute} \\ &= 1 - (1-\hat{p}_s)(1-\hat{p}_a) \text{ from (A.4) and (A.5),} \end{aligned}$$

and the result is proved.

For the binomial parameter, p_a, a sufficient statistic r is the number of nonconforming items in the sample of size n_a. From (A.5) follows the well-known result that the MVU estimator of p_a is

$$\hat{p}_a = r/n_a. \tag{A.6}$$

Denote the sample values on the variable by $x_1, x_2, ..., x_{n_s}$. For the parameter p_v, one pair of sufficient statistics is $t = (\bar{x}, S)$ where

$$\bar{x} = \sum_{i=1}^{n_s} x_i/n_s$$

and

$$S^2 = \sum_{i=1}^{n_s} (x_i - \bar{x})^2$$

Bowker and Goode (1952) showed that the MVU estimator of p_v for combined double specification limits (L, U) is, in the present notation,

$$\hat{p}_s = B(\tfrac{1}{2}\max[0, \ 1-\{(U-\bar{x})\sqrt{n_s}\}/\{S\sqrt{(n_s-1)}\}])$$

$$+ \ B(\tfrac{1}{2}\max[0, \ 1-\{(\bar{x}-L)\sqrt{n_s}\}/\{S\sqrt{(n_s-1)}\}]) \tag{A.7}$$

where $B(.)$ denotes the distribution function of the symmetric beta distribution of the first kind with parameters $\tfrac{1}{2}n_s - 1$.

APPENDIX B: THE DISTRIBUTION FUNCTION OF $\hat{p}_s$ FOR A SINGLE NORMALLY DISTRIBUTED VARIABLE WITH A SINGLE SPECIFICATION LIMIT

Consider a normally distributed variable with unknown mean and variance. Without loss of generality, suppose that the single specification limit is a lower one, L. Setting $U = \infty$ in (6) we have

$$\hat{p}_s = B(\tfrac{1}{2}\max[0, \ 1-\{(\bar{x}-L)\sqrt{n_s}\}/\{S\sqrt{(n_s-1)}\}]). \tag{B.1}$$

The process fraction nonconforming, p, is the tail area to the left of the value L under a normal distribution with mean μ and standard deviation σ. It follows that the relationship between p, L, μ and σ is given by

$$p = \Phi\{(L-\mu)/\sigma\} \tag{B.2}$$

where Φ is the standard normal distribution function. Since $\Phi(-x) = 1 - \Phi(x)$, (B.2) may be rewritten

$$1-p = \Phi\{(\mu-L)/\sigma\}. \tag{B.3}$$

Applying the inverse normal transformation Φ^{-1} to both sides of this equation gives $K_p = (\mu-L)/\sigma$, i.e.

$$\mu = L + \sigma K_p \tag{B.4}$$

where K_p is the upper p-fractile of the standard normal distribution. Thus the probability of accepting each lot on original inspection when the process fraction nonconforming is p is

$$P(\hat{p} \le p^* \mid p) = P\left\{B(\tfrac{1}{2}\max[\,0,\ 1-\{(\overline{x}-L)\sqrt{n_s}/\{S\sqrt{(n_s-1)}\}\,]) \le p^* \mid p\right\}. \tag{B.5}$$

Applying the inverse beta transformation B^{-1} to both sides of the inequality on the right hand side of expression (B.5) gives

$$P(\hat{p} \le p^* \mid p) = P\left(\tfrac{1}{2} - \tfrac{1}{2}\{(\overline{x}-L)\sqrt{n_s}/\{S\sqrt{(n_s-1)}\} \le \beta_{p^*} \mid p\right)$$

$$= P\left(\{(\overline{x}-L)\sqrt{n_s}/\{S\sqrt{(n_s-1)}\} \ge 1-2\beta_{p^*} \mid p\right). \tag{B.6}$$

But $\overline{x} - L = \overline{x} - \mu + \sigma K_p$ from (B.4). Substituting this result into (B.6), multiplying both sides of the inequality by (n_s-1) and rearranging gives

$$P(\hat{p} \le p^* \mid p) = P\left(\{(\overline{x}-\mu)/(\sigma/\sqrt{n_s}) + \sqrt{n_s}K_p\}/(s/\sigma) \ge (n_s-1)(1-2\beta_{p^*})\right)$$

where $s = S/\sqrt{(n_s-1)}$ is the sample standard deviation. As $(\overline{x}-\mu)/(\sigma/\sqrt{n_s})$ is a standard normal variate and s/σ is a $\chi_{n-1}/\sqrt{(n_s-1)}$ variate, it follows that

$$P(\hat{p} \le p^* \mid p) = P\left\{t_{n-1, \sqrt{n_s}K_p} \ge (n_s-1)(1-2\beta_{p^*})\right\} \tag{B.7}$$

where $t_{\nu,\delta}$ is a non-central t variate with degrees of freedom ν and eccentricity parameter δ.

APPENDIX C : THE SUBSCRIPT OF THE LAST NON-NEGATIVE ELEMENT OF $\{p_r^*\}$

The function p_r^* defined by

$$p_r^* = \begin{cases} p^* - r(1-p^*)/(n_a-r) & \text{for } r = 0, 1, ..., n_a - n_s \\ p^* - r(1-p^*)/(n_s-1) & \text{for } r = n_a-n_s+1, n_a-n_s+2, ... \end{cases} \tag{C.1}$$

is the acceptance constant for sampling by variables in "s" method dependent attriables inspection.

Now $\{p_r^*\}$ is clearly a strictly monotonic decreasing sequence for $r = 0, 1, ..., n_a - n_s$ and for $r = n_a-n_s+1,\ n_a-n_s+2, ...$ Furthermore,

$$p_{n_a-n_s}^* - p_{n_a-n_s+1}^* = (n_a-n_s+1)(1-p^*)/(n_s-1) - (n_a-n_s)(1-p^*)/n_s$$

$$= n_a(1-p^*)/\{n_s(n_s-1)\}$$

which is always positive. It follows that $\{p_r^*\}$ is a strictly monotonic decreasing sequence for $r = 0, 1, 2, ...$

Denote by c the subscript of the last non-negative element of this sequence. Thus c is such that $p_c \ge 0$ and $p_{c+1} < 0$. To find c consider first the case where $p_{n_a-n_s+1}^* < 0$. It follows using (C.1) that c is the integral part of the solution in r to $(n_a-r)p^* = r(1-p^*)$, i.e. $c = \lfloor n_a p^* \rfloor$ provided $c \le n_a-n_s$, i.e. provided $n_s \le n_a - \lfloor n_a p^* \rfloor$. Consider now the case where $p_{n_a-n_s+1}^* \ge 0$. Here c is the integral part of the solution in r to $p^* = r(1-p^*)/(n_s-1)$, i.e. $c = \lfloor (n_s-1)p^*/(1-p^*) \rfloor$ provided $c \ge n_a-n_s + 1$, i.e. provided $(n_s-1)p^* \ge \{(n_a-n_s+1)\cdot(1-p^*)\}$, i.e provided $n_s \ge n_a+1-n_a p^*$, i.e. provided $n_s \ge n_a+1-\lfloor n_a p^* \rfloor$.

In summary, the subscript of the last non-negative value of $\{p_r^*\}$ is

$$c = \begin{cases} \lfloor n_a p^* \rfloor & \text{when } n_s \le n_a - \lfloor n_a p^* \rfloor, \\ \lceil (n_s-1)p^*/(1-p^*) \rceil & \text{otherwise.} \end{cases} \tag{C.2}$$

Sensitivity to Process Quality under the Proposed Revision of ISO 2859

T. Koyama, Shido – cho, Kagawa, Japan

1. INTRODUCTION

ISO TC69/SC5/WG2 proposed the following major modifications of ISO 2859 [1] sampling system ;

a) the modified switching rule from normal to reduced inspection and that from tightened inspection to discontinuation of inspection ;

b) the proposed fractional acceptance number plans as an optional procedure, instead of arrows between the acceptance numbers of 0 and 1, which are essentially the same as chain sampling plans by Dodge [2].

It was necessary for the members of the working group to recognize average run lengths (ARLs), average acceptance probabilities (AAPs) and average sample sizes (ASSs) to introduce the modifications. For this purpose, first, the author derived sequential and average probabilities on normal, tightened, reduced inspection and switching probabilities to discontinuation of inspection, examining sensitivity to process changes. Second, he analyzed ARLs, AAPs and ASSs with expected switching probabilities (ESPs) ;

1) by calculating switching probabilities at the t – th lot of inspection directly, using transition probabilities ;

2) from generating functions obtained by Mason's reduction method [3] from signal flow graphs (SFGs) composed of both transition probabilities and a time delay operator ; and

3) from generating functions in the form of a determinant from SFGs in 2).

Each of these three methods has strong points and weak points about difficulty in calculation, computation time, and time series analysis.

Abbreviations and symbols are summarized in Appendix.

2. MAJOR MODIFICATIONS IN THE PROPOSED REVISION OF ISO 2859

2.1 The switching rule from normal to reduced inspection

The proposed revision [4, 5] of ISO 2859 includes the modified switching rule from normal to reduced inspection, as follows :

Reduced inspection may be instituted after 10 successive lots (or 16 successive lots for $Ac_N = 0$) have been accepted under normal inspection provided that

a) these lots would have been acceptable if AQL had been one step tighter ;

b) production is in statistical control ;

c) reduced inspection is considered desirable by responsible authority.

Frontiers in Statistical Quality Control 4
Ed. by Lenz et al.
© Physica-Verlag Heidelberg 1992

This switching rule is similar to that specified in ISO 3951 [6]. A simpler procedure will result in practice than the switching procedure specified in ISO 2859 or MIL − STD − 105D [7] including limit numbers.

2.2 The switching rule from tightened inspection to discontinuation of inspection

If the cumulative number of rejected lots on tightened inspection reach 5, sampling inspection is to be discontinued pending action. This switching rule has been specified in ISO 2859 − 1 [5] and is identical to that in JIS Z 9015 [8].

2.3 Fractional acceptance number plans

Fractional acceptance number plans in an optional procedure in the proposed revision of ISO 2859 − 1 may enable us not to use arrows between the acceptance numbers of 0 and 1 in the sampling inspection plans. An example when AQL = 1.0 % on normal inspection is shown in Table 1.

Table 1: Comparison of Ac_N at AQL = 1.0 %

code letter	n	ISO 2859 − 1	optional proce − dure AQL = 1.0 %	optional proce − dure AQL = 0.65 %
E	13	0	0	↓
F	20	↑	1/3	0
G	32	↓	1/2	1/3
H	50	1	1	1/2
J	80	2	2	1

3. HOW TO OBTAIN ARL, AAP AND ASS

3.1 Direct calculation with transition probabilities

The switching scheme combined with the switching rules in the proposed revision of ISO 2859 − 1 can be expressed by a transition probability matrix of Equation (1) and a SFG in Figure 1, when $Ac_N \geqq 2$, $Ac_T \geqq 1$ and $Ac_R \geqq 1$.

If we replace P_i (i = N, R, T, L, M) and Q_i (i = N, R, T) in Equation (1) with P_i (t) (i = N, R, T, L, M) and Q_i (t) (i = N, R, T), we can obtain a transition probability matrix P_t.

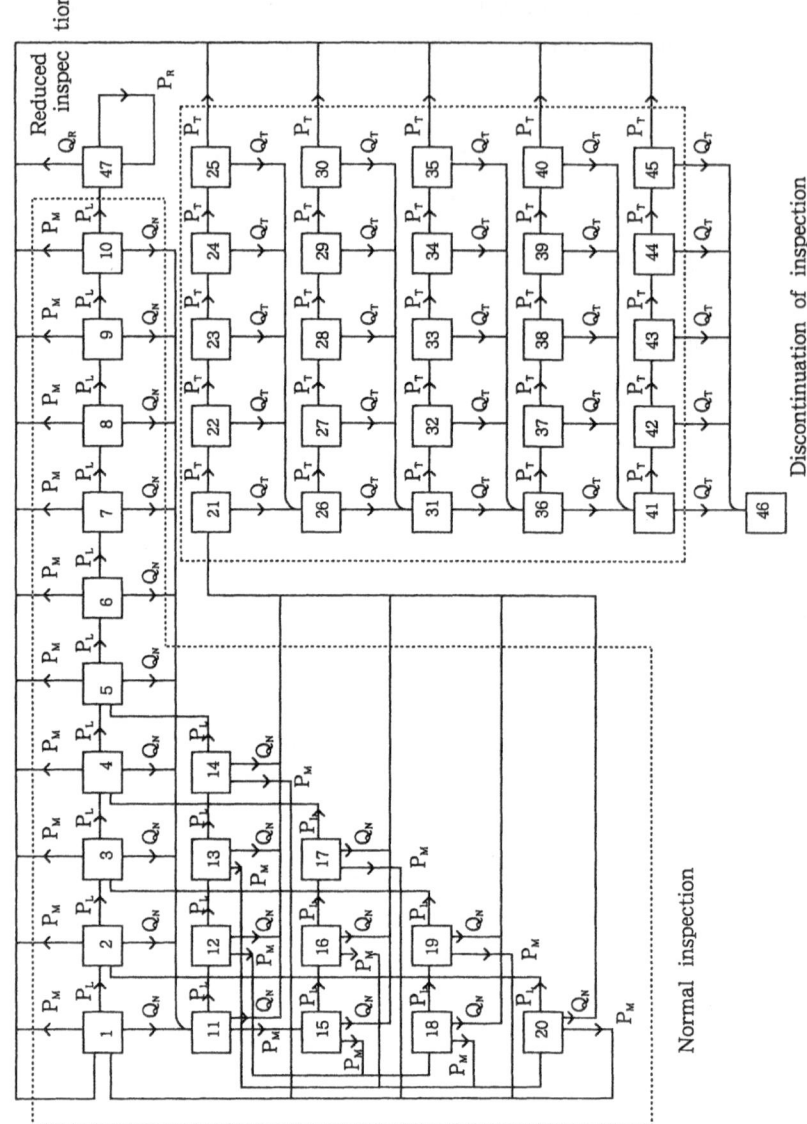

Figure 1: A SFG for switching from normal inspection to discontinuation of inspection ; $Ac_N \geqq 2$

$$
P = \begin{bmatrix}
P_N & P_L & 0 & 0 & 0 & 0 & 0 & 0 & 0 & 0 & Q_N & . & 0 & 0 & 0 & 0 & 0 & 0 & . & 0 \\
P_N & 0 & P_L & 0 & 0 & 0 & 0 & 0 & 0 & 0 & Q_N & . & 0 & 0 & 0 & 0 & 0 & 0 & . & 0 \\
P_N & 0 & 0 & P_L & 0 & 0 & 0 & 0 & 0 & 0 & Q_N & . & 0 & 0 & 0 & 0 & 0 & 0 & . & 0 \\
P_N & 0 & 0 & 0 & P_L & 0 & 0 & 0 & 0 & 0 & Q_N & . & 0 & 0 & 0 & 0 & 0 & 0 & . & 0 \\
P_N & 0 & 0 & 0 & 0 & P_L & 0 & 0 & 0 & 0 & Q_N & . & 0 & 0 & 0 & 0 & 0 & 0 & . & 0 \\
P_N & 0 & 0 & 0 & 0 & 0 & P_L & 0 & 0 & 0 & Q_N & . & 0 & 0 & 0 & 0 & 0 & 0 & . & 0 \\
P_N & 0 & 0 & 0 & 0 & 0 & 0 & P_L & 0 & 0 & Q_N & . & 0 & 0 & 0 & 0 & 0 & 0 & . & 0 \\
P_N & 0 & 0 & 0 & 0 & 0 & 0 & 0 & P_L & 0 & Q_N & . & 0 & 0 & 0 & 0 & 0 & 0 & . & 0 \\
P_N & 0 & 0 & 0 & 0 & 0 & 0 & 0 & 0 & P_L & Q_N & . & 0 & 0 & 0 & 0 & 0 & 0 & . & 0 \\
P_N & 0 & 0 & 0 & 0 & 0 & 0 & 0 & 0 & 0 & Q_N & . & 0 & 0 & 0 & 0 & 0 & 0 & . & P_L \\
0 & 0 & 0 & 0 & 0 & 0 & 0 & 0 & 0 & 0 & 0 & . & Q_N & 0 & 0 & 0 & 0 & 0 & . & 0 \\
. & . & . & . & . & . & . & . & . & . & . & & . & . & . & . & . & . & & . \\
0 & 0 & 0 & 0 & 0 & 0 & 0 & 0 & 0 & 0 & 0 & . & 0 & P_T & 0 & 0 & 0 & Q_T & . & 0 \\
0 & 0 & 0 & 0 & 0 & 0 & 0 & 0 & 0 & 0 & 0 & . & 0 & 0 & P_T & 0 & 0 & Q_T & . & 0 \\
0 & 0 & 0 & 0 & 0 & 0 & 0 & 0 & 0 & 0 & 0 & . & 0 & 0 & 0 & P_T & 0 & Q_T & . & 0 \\
0 & 0 & 0 & 0 & 0 & 0 & 0 & 0 & 0 & 0 & 0 & . & 0 & 0 & 0 & 0 & P_T & Q_T & . & 0 \\
P_T & 0 & 0 & 0 & 0 & 0 & 0 & 0 & 0 & 0 & 0 & . & 0 & 0 & 0 & 0 & 0 & Q_T & . & 0 \\
0 & 0 & 0 & 0 & 0 & 0 & 0 & 0 & 0 & 0 & 0 & . & 0 & 0 & 0 & 0 & 0 & 0 & . & 0 \\
. & . & . & . & . & . & . & . & . & . & . & & . & . & . & . & . & . & & . \\
Q_R & 0 & 0 & 0 & 0 & 0 & 0 & 0 & 0 & 0 & 0 & . & 0 & 0 & 0 & 0 & 0 & 0 & . & P_R
\end{bmatrix}
\qquad (1)
$$

1 2 3 4 5 6 7 8 9 10 11 21 22 23 24 25 26 47

We can calculate the probability distribution at the $(t+1)$ − th lot of inspection from

$$ p^{(t+1)} = p^{(t)} P_t . \qquad (2) $$

Equation (2) is equal to

$$ [\phi(j, t+1)] = [\phi(i, t)] [p_{ij}(t)] \quad . \qquad (3) $$

Let $p^{(1)}$ be a initial probability vector, then Equation (4) will be used for analyses of switching scheme in the proposed revision of ISO 2859 when normal inspection is applied at the start of inspection ;

$$ p^{(1)} = [\phi(1,1)=1, \phi(i, 1)=0 \, (i=2\sim47)] . \qquad (4) $$

Although Equations (2) and (3) are useful for theoretical analyses, Equation (5) was used for numerical analyses so as to save computation times with an electronic computer,

$$ \phi(j, t+1) = \sum_i \phi(i, t) \, p_{ij}(t) \quad . \qquad (5) $$

The reason is because P_t is a sparse matrix comparable to Equation (1). From Equation (5), we obtain

$$ \phi_N(t+1) = \sum_{i=1}^{20} \phi(j, t+1) \quad , \qquad (6) $$

$$ \phi_T(t+1) = \sum_{i=21}^{45} \phi(j, t+1) \quad , \qquad (7) $$

$$ \phi_R(t+1) = \phi(47, t+1) \quad , \qquad (8) $$

$$ \phi_D(t+1) = \phi(46, t+1) \quad . \qquad (9) $$

If we use Equation (1), ARL from normal inspection to discontinuation of inspection, ARL_{NRT-D}, is

$$ARL_{NRT\text{-}D} = \sum_{t=1}^{\infty} (t-1)\,\phi_D(t) \tag{10}$$

From Equations (6) $\sim$ (8)

$$Frac_N(t) = \phi_N(t)/[\phi_N(t) + \phi_T(t) + \phi_R(t)] \tag{11}$$

$$Frac_N(t) = \phi_T(t)/[\phi_N(t) + \phi_T(t) + \phi_R(t)] \tag{12}$$

$$Frac_R(t) = \phi_R(t)/[\phi_N(t) + \phi_T(t) + \phi_R(t)] \tag{13}$$

Using Equatios (11) $\sim$ (13), we obtain

$$AAP(t) = P_N(t)\,Frac_N(t) + P_T(t)\,Frac_T(t) + P_R(t)\,Frac_R(t) \;, \tag{14}$$

$$ASS(t) = n_N\,Frac_N(t) + n_T\,Frac_T(t) + n_R\,Frac_R(t) \;. \tag{15}$$

For constant process average, the overall acceptance probability and the overall sample size are given by

$$AAP = \sum_{t=1}^{\infty} [\phi_N(t) + \phi_T(t) + \phi_R(t)]\,AAP(t)\,/\,\sum_{t=1}^{\infty} [\phi_N(t) + \phi_T(t) + \phi_R(t)\} \;, \tag{16}$$

$$ASS = \sum_{t=1}^{\infty} [\phi_N(t) + \phi_T(t) + \phi_R(t)]\,ASS(t)\,/\,\sum_{t=1}^{\infty} [\phi_N(t) + \phi_T(t) + \phi_R(t)] \;. \tag{17}$$

3.2 Generating functions

A SFG induced by attaching a time delay operator $x^{9)}$ to each branch of the probabilities of the SFG in Figure 1 can be reduced to a SFG in Figure 2 by Mason's method, where

$$G_{N\text{-}T}(x) = G_{1\text{-}21}(x) = Q_N^2 x^2 \left[\sum_{\nu=0}^{3} (P_N x)^{\nu} \right]\left[\sum_{\nu=0}^{9} (P_L x)^{\nu} \right] / \Delta(x) \;, \tag{18}$$

$$G_{N\text{-}R}(x) = G_{1\text{-}47}(x) = P_L^{10} x^{10} \left[1 + Q_N x \sum_{\nu=0}^{3} (P_N x)^{\nu} \right] / \Delta(x) \tag{19}$$

$$\Delta(x) = 1 + Q_N x \sum_{\nu=0}^{3} (P_N x)^{\nu} - \left[(P_M + Q_N)x + P_M Q_N x^2 \sum_{\nu=0}^{3} (P_N x)^{\nu} \right] \sum_{\nu=0}^{9} (P_L x)^{\nu} \;, \tag{20}$$

where

$$G_{T\text{-}N}(x) = G_{21\text{-}1}(x) = G(x) \left[1 + H(x) + H^2(x) + H^3(x) + H^4(x) \right] \tag{21}$$

$$G_{T\text{-}D}(x) = G_{21\text{-}46}(x) = H^5(x) \;, \tag{22}$$

$$G(x) = P_T^5\, x^5 \;, \tag{23}$$

$$H(x) = Q_T x + Q_T P_T x^2 + Q_T P_T^2 x^3 + Q_T P_T^3 x^4 + Q_T P_T^4 x^5 \;, \tag{24}$$

and where

$$G_{R\text{-}N}(x) = G_{47\text{-}1}(x) = Q_R x / (1 - P_R x) \;. \tag{25}$$

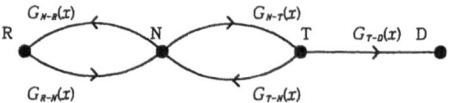

Figure 2: A reduced SFG for switching from normal inspection to discontinuation

Expected switching probability (ESP) from i to j was obtained by

$$S_{i\text{-}j} = G_{i\text{-}j}(1) \quad . \tag{26}$$

From Equations (18) $\sim$ (20) and (26)

$$S_{N\text{-}R} = G_{N\text{-}R}(1) = P_L^{10}(2 - P_N^4)/\Delta(1) \quad , \tag{27}$$

$$S_{N\text{-}T} = G_{N\text{-}T}(1) = Q_N(1 - P_N^4)(1 - P_L^{10})/[Q_L\Delta(1)] \quad , \tag{28}$$

where

$$\Delta(1) = P_L^{10}(2 - P_N^4) + Q_N(1 - P_N^4)(1 - P_L^{10})/Q_L \quad . \tag{29}$$

From Equations (21) $\sim$ (24) and (26)

$$S_{T\text{-}N} = G_{T\text{-}N}(1) = 1 - (1 - P_T^5)^5 \quad , \tag{30}$$

$$S_{T\text{-}D} = G_{T\text{-}D}(1) = (1 - P_T^5)^5 \quad . \tag{31}$$

From Equations (25) and (26)

$$S_{R\text{-}N} = G_{R\text{-}N}(1) = 1 \quad . \tag{32}$$

ARL from i to j is given by

$$ARL_{i\text{-}j} = G_{i\text{-}j}''(1) / G(1) \quad . \tag{33}$$

From Equations (18) $\sim$ (20) and (33)

$$ARL_{N\text{-}R} = 10 + \frac{1 - P_N^4}{2 - P_N^4} + \frac{P_N - 4P_N^4 + 3P_N^5}{Q_N(1 - P_N^4)} - \frac{\Delta'(1)}{\Delta(1)} \quad , \tag{34}$$

$$ARL_{N\text{-}T} = 2 + \frac{P_N - 4P_N^4 + 3P_N^5}{Q_N(1 - P_N^4)} + \frac{P_L - 10P_L^{10} + 9P_L^{11}}{Q_L(1 - P_L^{10})} - \frac{\Delta'(1)}{\Delta(1)} \quad , \tag{35}$$

where

$$\Delta'(1) = \frac{1 - 5P_N^4 + 4P_N^5}{Q_N} - \frac{1 - 11P_L^{10} + 10P_L^{11}}{Q_L} - \frac{P_M(1 - P_N^4)(2 - P_L - 12P_L^{10} + 11P_L^{11})}{Q_L^2}$$

$$- \frac{P_N P_M(1 - 4P_N^3 + 3P_N^4)(1 - P_L^{10})}{Q_N Q_L} \quad . \tag{36}$$

From Equations (21) $\sim$ (24) and (33),

$$ARL_{T\text{-}N} = 5 + H'(1)\frac{1 + 2H(1) + 3H^2(1) + 4H^3(1)}{1 + H(1) + H^2(1) + H^3(1) + H^4(1)} \quad , \tag{37}$$

$$ARL_{T\text{-}D} = 5H'(1)/H(1) \quad , \tag{38}$$

where

$$H(1) = 1 - P_T^5 \tag{39}$$

$$H'(1) = (1 - 6P_T^5 + 5P_T^6)/Q_T \quad . \tag{40}$$

From Equation (25), (26) and (33)

$$ARL_{R\text{-}N} = 1/Q_R \quad . \tag{41}$$

Using Mason's method, the SFG in Figure 2 can be reduced to a generating function from normal inspection to discontinuation of inspection, $G_{NRT\text{-}D}(x)$, as follows ;

$$G_{NRT\text{-}D}(x) = \frac{G_{N\text{-}T}(x) G_{T\text{-}D}(x)}{1 - G_{N\text{-}R}(x) G_{R\text{-}N}(x) - G_{N\text{-}T}(x) G_{T\text{-}N}(x)} \tag{42}$$

Since $G_{NRT-D}(1) = 1$,

$$ARL_{NRT-D} = G'_{NRT-D}(1)$$

$$= \frac{S_{N-R}\,(ARL_{N-R} + ARL_{R-N})}{S_{N-T}\,S_{T-D}} + \frac{ARL_{N-T} + S_{T-N}\,ARL_{T-N}}{S_{T-D}} + ARL_{T-D} \tag{43}$$

Equation (43) is identical to the result given by Thyregod [10] who has analyzed the switching scheme by applying a theory of reccurent event. Furthermore, AAP and ASS were derived by applying the result given by Thyregod, as follows ;

$$AAP = \left[P_R \frac{S_{N-R}}{S_{N-T}} ARL_{R-N} + P_N \Big(\frac{S_{N-R}}{S_{N-T}} ARL_{N-R} + ARL_{N-T} \Big) + P_T (S_{T-N} ARL_{T-N} + S_{T-D} ARL_{T-D}) \right] / \Delta , \tag{44}$$

$$ASS = \left[n_R \frac{S_{N-R}}{S_{N-T}} ARL_{R-N} + n_N \Big(\frac{S_{N-R}}{S_{N-T}} ARL_{N-R} + ARL_{N-T} \Big) + n_R (S_{T-N} ARL_{T-N} + S_{T-D} ARL_{T-D}) \right] / \Delta , \tag{45}$$

where

$$\Delta = \frac{S_{N-R}}{S_{N-T}} ARL_{R-N} + \frac{S_{N-R}}{S_{N-T}} ARL_{N-R} + ARL_{N-T} + S_{T-N} ARL_{T-N} + S_{T-D} ARL_{T-D} \quad . \tag{46}$$

When $Ac_N = 0$, switching rule from normal to reduced inspection changes from ten to sixteen successive accepted lots, so that, using the result by Koyama [11],

$$S_{N-R} = P_N^{16} (2 - P_N^4) / (1 - P_N^4 + P_N^{16}) , \tag{47}$$

$$S_{N-T} = (1 - P_N^4)(1 - P_N^{16}) / (1 - P_N^4 + P_N^{16}) , \tag{48}$$

$$ARL_{N-R} = 16 + \frac{1 - 5P_N^4 + 4P_N^5}{Q_N(2 - P_N^4)} + \frac{5P_N^4 - 4P_N^5 - 17P_N^{16} + 16P_N^{17}}{Q_N(1 - P_N^4 + P_N^{16})} , \tag{49}$$

$$ARL_{N-T} = 2 + \frac{P_N - 4P_N^4 + 3P_N^5}{Q_N(1 - P_N^4)} + \frac{P_N - 16P_N^{16} + 15P_N^{17}}{Q_N(1 - P_N^{16})} + \frac{5P_N^4 - 4P_N^5 - 17P_N^{16} + 16P_N^{17}}{Q_N(1 - P_N^4 + P_N^{16})} \quad . \tag{50}$$

Very short computation time is sufficient for numerical calculation of ESP, ARL, AAP and ASS obtained theoretically from the generating functions by Mason's method. However, if the number of loops in a SFG increases, derivation of generating functions turns difficult. To overcome this difficulty, we may use the following determinant method in practice.

3.3 Generating functions in the form of determinants

When we attach a branch of unit to the node 1 in the SFG in 3.2, considering node signals z_i , we can lead to simultaneous equations of regarding z_i. Generating functions are obtained by calculating z_i / z_0 [10] in the form of determinants with the Cramer's formula.

A simple example : $G_{R-N}(x)$ for $Ac_R = 1/2$

A SFG from reduced to normal inspection with the time delay operator is shown in Figure 3. From Figure 3, we obtain Equation (52), as

$$
\left.
\begin{aligned}
z_1 &= z_R \\
z_2 &= (P_{R0}\,x)\,z_1 + (P_{R0}\,x)\,z_2 + (P_{R0}\,x)\,z_3 \\
z_3 &= (P_{R1}\,x)\,z_2 \\
z_N &= (Q_{R1}\,x)\,z_1 + (Q_{R2}\,x)\,z_2 + (Q_{R1}\,x)\,z_3
\end{aligned}
\right\} \tag{51}
$$

Using the Cramer's formula, we obtain

$$
G_{R\text{-}N}(x) = \frac{z_N}{z_R} =
\begin{vmatrix}
-1 & 0 & 0 & -1 \\
P_{R0}x & P_{R0}x-1 & P_{R0}x & 0 \\
0 & P_{R1}x & -1 & 0 \\
Q_{R1}x & Q_{R1}x & Q_{R1}x & 0
\end{vmatrix}
\Bigg/
\begin{vmatrix}
-1 & 0 & 0 & 0 \\
P_{R0}x & P_{R0}x-1 & P_{R0}x & 0 \\
0 & P_{R1}x & -1 & 0 \\
Q_{R1}x & Q_{R2}x & Q_{R1}x & -1
\end{vmatrix}
\equiv \frac{\Delta_{R\text{-}N}(x)}{\Delta_R(x)} \,, \tag{52}
$$

so that, considering $G_{R\text{-}N}(1) = 1$,

$$
ARL_{R\text{-}N} = G'_{R\text{-}N}(1) = \frac{\Delta'_{R\text{-}N}(1)}{\Delta_{R\text{-}N}(1)} - \frac{\Delta'_R(1)}{\Delta_R(1)} \tag{53}
$$

where

$\Delta'_{R\text{-}N}(1)$

$$
=
\begin{vmatrix}
0 & 0 & 0 & 0 \\
P_{R0} & P_{R0}-1 & P_{R0} & 0 \\
0 & P_{R1} & -1 & 0 \\
Q_{R1} & Q_{R2} & Q_{R1} & 0
\end{vmatrix}
+
\begin{vmatrix}
-1 & 0 & 0 & -1 \\
P_{R0} & P_{R0} & P_{R0} & 0 \\
0 & P_{R1} & -1 & 0 \\
Q_{R1} & Q_{R2} & Q_{R1} & 0
\end{vmatrix}
+
\begin{vmatrix}
-1 & 0 & 0 & -1 \\
P_{R0} & P_{R0}-1 & P_{R0} & 0 \\
0 & P_{R1} & 0 & 0 \\
Q_{R1} & Q_{R2} & Q_{R1} & 0
\end{vmatrix}
+
\begin{vmatrix}
-1 & 0 & 0 & -1 \\
P_{R0} & P_{R0}-1 & P_{R0} & 0 \\
0 & P_{R1} & -1 & 0 \\
Q_{R1} & Q_{R2} & Q_{R1} & 0
\end{vmatrix}
\tag{54}
$$

$\Delta'_R(1)$

$$
=
\begin{vmatrix}
0 & 0 & 0 & 0 \\
P_{R0} & P_{R0}-1 & P_{R0} & 0 \\
0 & P_{R1} & -1 & 0 \\
Q_{R1} & Q_{R2} & Q_{R1} & -1
\end{vmatrix}
+
\begin{vmatrix}
-1 & 0 & 0 & 0 \\
P_{R0} & P_{R0} & P_{R0} & 0 \\
0 & P_{R1} & -1 & 0 \\
Q_{R1} & Q_{R2} & Q_{R1} & -1
\end{vmatrix}
+
\begin{vmatrix}
-1 & 0 & 0 & 0 \\
P_{R0} & P_{R0}-1 & P_{R0} & 0 \\
0 & P_{R1} & 0 & 0 \\
Q_{R1} & Q_{R2} & Q_{R1} & -1
\end{vmatrix}
+
\begin{vmatrix}
-1 & 0 & 0 & 0 \\
P_{R0} & P_{R0}-1 & P_{R0} & 0 \\
0 & P_{R1} & -1 & 0 \\
Q_{R1} & Q_{R2} & Q_{R1} & 0
\end{vmatrix}
\tag{55}
$$

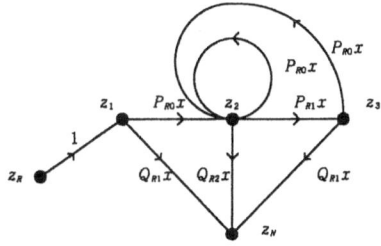

Figure 3: A SFG for switching from reduced to normal inspection ; $Ac_R = 1/2$

By the determinant method, the author calculated ESPs and ARLs for $Ac_N = 1/3, 1/2, 1$; and $Ac_R = 1/5, 1/3, 1/2$; as shown in Tables 4~7.

4. NUMERICAL RESULTS

4.1 Numerically calculated ESPs, ARLs and ASSs

Table 2 shows ESPs and ARLs for representative sets of switching scheme under the proposed revision of ISO 2859 − 1. Tables 3~9 show the numerically calculated results, namely, S_{N-R}, S_{N-T}, S_{T-D}, ARL_{N-R}, ARL_{N-T}, ARL_{T-D}, ARL_{R-N} and ARL_{NRT-D}.

Table 2: Sampling schemes used for calculating ESPs and ARLs

switching scheme	normal inspection			tightened inspection			reduced inspection			Table
	Ac_N	n_N^*	n_N AQL	Ac_T	n_T^*	n_T AQL	Ac_R	n_R^*	n_R AQL	
1	0	13	0.1262	0	20	0.2000	0	5	0.05024	3
2	1/3	20	0.2000	0	20	0.2000	1/5	8	0.07962	4
3	1/2	32	0.3170	1/3	32	0.3170	1/3	13	0.1262	5
4	1	50	0.5024	1/2	50	0.5024	1/2	20	0.2000	6
5	1	50	0.5024	1	80	0.7962	1	32	0.3170	7
6	2	80	0.7962	1	80	0.7962	1	32	0.3170	8
7	7	80	0.7962	5	80	0.7962	4	32	0.3170	9

* sample sizes used for calculating ASSs

4.2 Sensitivity to process quality

The author made the process average changed periodically to examine influence on the probabilities of $\phi_N(t)$, $\phi_R(t)$, $\phi_T(t)$ and $\phi_D(t)$ so as to recognize the changes under the proposed revision of ISO 2859 − 1. For a sampling scheme of $(Ac_N = 2, Ac_T = 1, Ac_R = 1)$ and for that of $(Ac_N = 1/2, Ac_T = 1/3, Ac_R = 1/3)$, the author examined sensitivity to process changes between p/AQL = 1.0 and p/AQL = 2.0 every 5, 10, 20 and 50 lots of inspection. For $Ac_N = 2$, $Ac_T = 1$ and $Ac_R = 1$, Figure 4 shows that $Frac_N$ (t), $Frac_T$ (t) and $Frac_R$ (t) reach stable state about 30 lots after the start of inspection, while for $Ac_N = 1/2$, $Ac_T = 1/3$ and $Ac_R = 1/3$ about 40 lots in Figure 9, if no process change occurs. Despite that the process changes every 5 lots of inspection, $\phi_R(t)$ is affected very little as shown in Figures 5 and 10. When the process changes every 10 lots of inspection, Figures 6 and 11 show that sensitivity turns a little larger but insufficient to detect the process change. When the process changes every 20 lots of inspection, Figures 7 and 12 show sufficient sensitivity to the process changes if we pay attention to the switches between normal and tightened inspection, but insufficient sensitivity if we watch the switches between normal and reduced inspection. The reason is that the minimum number of lots for switching from normal to reduced inspection is 10 lots, while 2 lots are sufficient for switching from normal to tightened inspection and 5 lots for switching back to normal inspection.

Consequently, we may check the process condition, if possible, when the number of lots on

Table 3: Switching characteristics

$Ac_N = 0$, $Ac_T = 0$, $Ac_R = 0$ $L_R = 16$

$n_N \cdot AQL = 0.1262$, $n_T \cdot AQL = 0.2000$, $n_H \cdot AQL = 0.05024$

p/AQL	$S_{N-R}^{2)}$	$S_{N-T}^{2)}$	$S_{T-D}^{2)}$	$ARL_{N-R}^{2)}$	$ARL_{R-N}^{2)}$	$ARL_{N-T}^{2)}$	$ARL_{T-D}^{2)}$	$ARL_{T-N}^{2)}$	$ARL_{NRT-D}^{2)}$
1.0	0.35035	0.64965	0.10093	24.7193	20.4086	16.3823	13.0339	8.0161	487.9
1.5	0.12782	0.87218	0.28297	22.8970	13.7759	13.3619	12.1111	8.6538	100.3
2.0	0.04413	0.95587	0.48332	21.0299	10.4606	10.5648	11.2533	8.8527	45.6
2.5	0.01524	0.98476	0.65165	19.7207	8.4722	8.5457	10.4718	8.8311	29.0
3.0	0.00532	0.99468	0.77465	18.8551	7.1474	7.1392	9.7720	8.7094	21.7
4.0	0.00067	0.99933	0.91172	17.8789	5.4928	5.4200	8.6134	8.3817	15.4
5.0	0.00009	0.99991	0.96676	17.3908	4.5018	4.4587	7.7403	8.0752	12.6

2) calculated with the theoretical equations in 3.2   3) calculated with the generating functions in 3.3

Table 4: Switching characteristics

$Ac_N = 1/3$, $Ac_T = 0$, $Ac_R = 1/5$

$n_N \cdot AQL = 0.2000$, $n_T \cdot AQL = 0.2000$, $n_R \cdot AQL = 0.07962$

p/AQL	$S_{N-R}^{3)}$	$S_{N-T}^{3)}$	$S_{T-D}^{2)}$	$ARL_{N-R}^{3)}$	$ARL_{R-N}^{3)}$	$ARL_{N-T}^{3)}$	$ARL_{T-D}^{2)}$	$ARL_{T-N}^{2)}$	$ARL_{NRT-D}^{2)}$
1.0	0.56737	0.43263	0.10093	22.9223	43.3643	16.1622	13.0339	8.0161	1105.9
1.5	0.21833	0.78167	0.28297	21.5077	21.3482	14.5604	12.1111	8.6538	127.8
2.0	0.06943	0.93057	0.48332	17.9268	13.2653	10.7918	11.2533	8.8527	47.9
2.5	0.02186	0.97814	0.65165	15.3308	9.3531	8.0299	10.4718	8.8311	28.4
3.0	0.00710	0.99290	0.77465	13.7244	7.1367	6.2815	9.7720	8.7094	20.6
4.0	0.00082	0.99918	0.91172	12.0882	4.8087	4.4204	8.6134	8.3817	14.3
5.0	0.00010	0.99990	0.96676	11.3699	3.6494	3.5331	7.7403	8.0752	11.7

2) calculated with the theoretical equations in 3.2   3) calculated with the generating functions in 3.3

Table 5: Switching characteristics

$Ac_N = 1/2$, $Ac_T = 1/3$, $Ac_R = 1/3$

$n_N \cdot AQL = 0.3170$, $n_T \cdot AQL = 0.3170$, $n_R \cdot AQL = 0.1262$

p/AQL	$S_{N-R}^{3)}$	$S_{N-T}^{3)}$	$S_{T-D}^{2)}$	$ARL_{N-R}^{3)}$	$ARL_{R-N}^{3)}$	$ARL_{N-T}^{3)}$	$ARL_{T-D}^{2)}$	$ARL_{T-N}^{2)}$	$ARL_{NRT-D}^{2)}$	$ARL_{NRT-D}^{1)}$
1.0	0.57624	0.42376	0.077855	14.9547	31.1174	9.6815	10.232	7.208	1024.7	1024.7
1.5	0.22459	0.77541	0.27934	14.8531	15.7626	8.9942	9.859	7.965	94.4	94.4
2.0	0.06729	0.93272	0.52208	13.4656	9.5515	7.0536	9.258	8.225	33.5	33.5
2.5	0.01839	0.98161	0.71756	12.3289	6.7508	5.4787	8.610	8.216	20.0	20.0
3.0	0.00486	0.99514	0.84480	11.6057	5.1656	4.4321	8.000	8.092	14.8	14.8
4.0	0.00032	0.99968	0.95833	10.8775	3.5045	3.2993	7.014	7.782	10.8	10.8
5.0	0.00002	0.99998	0.98964	10.5668	2.6817	2.7656	6.344	7.532	9.2	9.2

1) directly calculated with transition probabilities   2) calculated with the theoretical equations in 3.2
3) calculated with the generating functions in 3.3

Table 6: Switching characteristics

$Ac_N = 1$, $Ac_T = 1/2$, $Ac_R = 1/2$

$n_N \cdot AQL = 0.5024$, $n_T \cdot AQL = 0.5024$, $n_R \cdot AQL = 0.2000$

p/AQL	$S_{N-R}^{3)}$	$S_{N-T}^{3)}$	$S_{T-D}^{2)}$	$ARL_{N-R}^{3)}$	$ARL_{R-N}^{3)}$	$ARL_{N-T}^{3)}$	$ARL_{T-D}^{2)}$	$ARL_{T-N}^{2)}$	$ARL_{NRT-D}^{2)}$
1.0	0.55558	0.44442	0.16156	22.1847	21.1840	17.9488	10.3789	7.7188	497.1
1.5	0.11215	0.88785	0.49249	19.4653	10.5777	13.9522	9.4943	8.2644	54.0
2.0	0.01459	0.98541	0.76188	15.2459	6.6690	8.8301	8.4900	8.2115	23.1
2.5	0.00175	0.99825	0.90322	13.1424	4.7726	6.1457	7.6026	7.9791	15.3
3.0	0.00020	0.99980	0.96361	12.0905	3.6972	4.7327	6.9012	7.7400	12.1
4.0	0.00000	1.00000	0.99543	11.1374	2.5697	3.3980	6.0015	7.3984	9.4
5.0	0.00000	1.00000	0.99947	10.7172	2.0129	2.8059	5.5383	7.2151	8.3

2) calculated with the theoretical equations in 3.2   3) calculated with the generating functions in 3.3

Table 7: Switching characteristics

$Ac_N = 1$, $Ac_T = 1$, $Ac_R = 1$

$n_N \cdot AQL = 0.5024$, $n_T \cdot AQL = 0.7962$, $n_R \cdot AQL = 0.3170$

p/AQL	$S_{N\text{-}R}^{3)}$	$S_{N\text{-}T}^{3)}$	$S_{T\text{-}D}^{2)}$	$ARL_{N\text{-}R}^{3)}$	$ARL_{R\text{-}N}^{3)}$	$ARL_{N\text{-}T}^{3)}$	$ARL_{T\text{-}D}^{2)}$	$ARL_{T\text{-}N}^{2)}$	$ARL_{NRT\text{-}D}^{2)}$
1.0	0.55558	0.44442	0.11691	22.1847	21.5170	17.9488	12.9341	8.1126	695.0
1.5	0.11215	0.88785	0.49915	19.4653	12.0675	13.9522	11.1843	8.8572	56.0
2.0	0.01459	0.98541	0.81198	15.2495	8.8324	8.8301	9.5162	8.6480	22.8
2.5	0.00175	0.99825	0.94435	13.1424	5.3046	6.1457	8.1663	8.2291	15.2
3.0	0.00020	0.99980	0.98555	12.0905	4.0614	4.7327	7.1833	7.8650	12.1
4.0	0.00000	1.00000	0.99922	11.1374	2.7640	3.3980	6.0435	7.4170	9.4
5.0	0.00000	1.00000	0.99997	10.7172	2.1267	2.8059	5.5124	7.2049	8.3

2) calculated with the theoretical equations in 3.2   3) calculated with the generating functions in 3.3

Table 8: Switching characteristics

$Ac_N = 2$, $Ac_T = 1$, $Ac_R = 1$

$n_N \cdot AQL = 0.7962$, $n_T \cdot AQL = 0.7962$, $n_R \cdot AQL = 0.3170$

p/AQL	$S_{N\text{-}R}^{2)}$	$S_{N\text{-}T}^{2)}$	$S_{T\text{-}D}^{2)}$	$ARL_{N\text{-}R}^{2)}$	$ARL_{R\text{-}N}^{2)}$	$ARL_{N\text{-}T}^{2)}$	$ARL_{T\text{-}D}^{2)}$	$ARL_{T\text{-}N}^{2)}$	$ARL_{NRT\text{-}D}^{2)}$	$ARL_{NRT\text{-}D}^{1)}$
1.0	0.79065	0.20935	0.11691	30.8712	21.5170	26.8287	12.9341	8.1126	2093.0	2093.0
1.5	0.14454	0.85546	0.49915	29.8936	12.0675	24.3795	11.1843	8.8572	83.1	83.1
2.0	0.00950	0.99050	0.81198	18.5383	8.8324	11.9976	9.5162	8.6480	26.6	26.6
2.5	0.00054	0.99946	0.94435	14.2420	5.3046	7.0707	8.1663	8.2291	16.1	16.1
3.0	0.00003	0.99997	0.98555	12.5291	4.0614	4.9622	7.1833	7.8650	12.3	12.3
4.0	0.00000	1.00000	0.99922	11.2933	2.7640	3.2776	6.0435	7.4170	9.3	9.3
5.0	0.00000	1.00000	0.99997	10.8836	2.1267	2.6392	5.5124	7.2049	8.2	8.2

1) directly calculated with transition probabilities   2) calculated with the theoretical equations in 3.2

Table 9: Switching characteristics

$Ac_N = 7$, $Ac_T = 5$, $Ac_R = 4$

$n_N \cdot AQL = 0.7962$, $n_T \cdot AQL = 0.7962$, $n_R \cdot AQL = 0.3170$

p/AQL	$S_{N-R}^{2)}$	$S_{N-T}^{2)}$	$S_{T-D}^{2)}$	$ARL_{N-R}^{2)}$	$ARL_{R-N}^{2)}$	$ARL_{N-T}^{2)}$	$ARL_{T-D}^{2)}$	$ARL_{T-N}^{2)}$	$ARL_{NRT-D}^{2)}$
1.0	0.98246	0.01754	0.01247	18.6693	105.4964	15.6323	13.9293	6.8093	559406.8
1.5	0.15351	0.84649	0.51538	33.4017	22.9806	27.8965	11.1129	8.8603	93.4
2.0	0.00040	0.99960	0.95416	14.7670	8.9245	7.5814	7.9977	8.1691	16.3
2.5	0.00000	1.00000	0.99847	11.7158	4.7299	3.8179	6.2283	7.4906	10.1
3.0	0.00000	1.00000	0.99997	10.9672	3.0355	2.7379	5.4826	7.1930	8.2
4.0	0.00000	1.00000	1.00000	10.5888	1.7609	2.1361	5.0669	7.0268	7.2
5.0	0.00000	1.00000	1.00000	10.5150	1.3257	2.0221	5.0077	7.0031	7.0

2) calculated with the theoretical equations in 3.2

Table 10: Switching characteristics under the switching rule from tightened inspection in ISO 2859

$Ac_N = 2$, $Ac_T = 1$, $Ac_R = 1$

$n_N \cdot AQL = 0.7962$, $n_T \cdot AQL = 0.7962$, $n_R \cdot AQL = 0.3170$

p/AQL	$S_{N-R}^{2)}$	$S_{N-T}^{2)}$	$S_{T-D}^{*2)}$	$ARL_{N-R}^{2)}$	$ARL_{R-N}^{2)}$	$ARL_{N-T}^{2)}$	$ARL_{T-D}^{*2)}$	$ARL_{T-N}^{*2)}$	$ARL_{NRT-D}^{*2)}$	ARL_{NRT-D}
1.0	0.79065	0.20935	0.31969	30.8712	21.5170	26.8287	10	6.4609	726.6	2093.0
1.5	0.14454	0.85546	0.75157	29.8936	12.0675	24.3795	10	6.8792	54.1	83.1
2.0	0.00950	0.99050	0.86279	18.5383	8.8324	11.9976	10	7.1080	25.3	26.6
2.5	0.00054	0.99946	0.93633	14.2420	5.3046	7.0707	10	7.2419	18.1	16.1
3.0	0.00003	0.99997	0.98708	12.5291	4.0614	4.9622	10	7.3251	15.1	12.3
4.0	0.00000	1.00000	0.99920	11.2933	2.7640	3.2776	10	7.4157	13.3	9.3
5.0	0.00000	1.00000	0.99996	10.8836	2.1267	2.6392	10	7.4580	12.6	8.2

2) calculated with the theoretical equations in 3.2

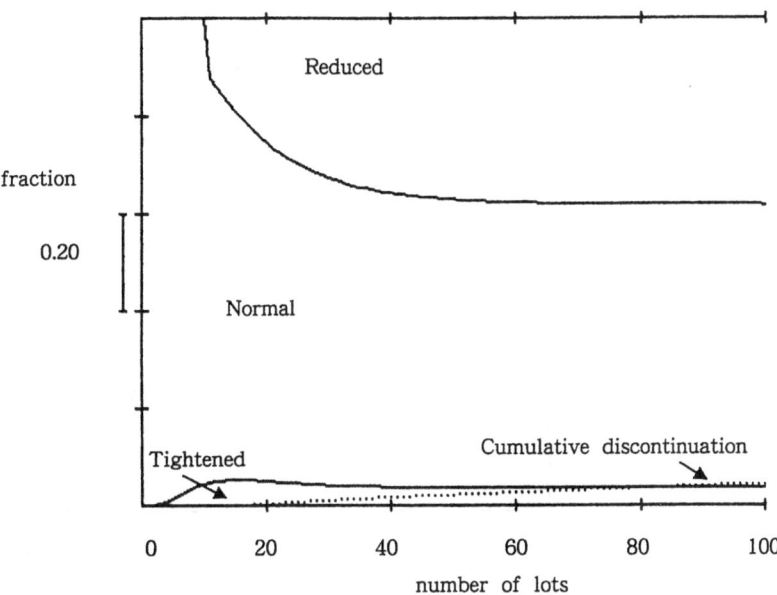

Figure 4: Fraction of inspection for p/AQL = 1.0 ; $Ac_N = 2$, $Ac_T = 1$ and $Ac_R = 1$

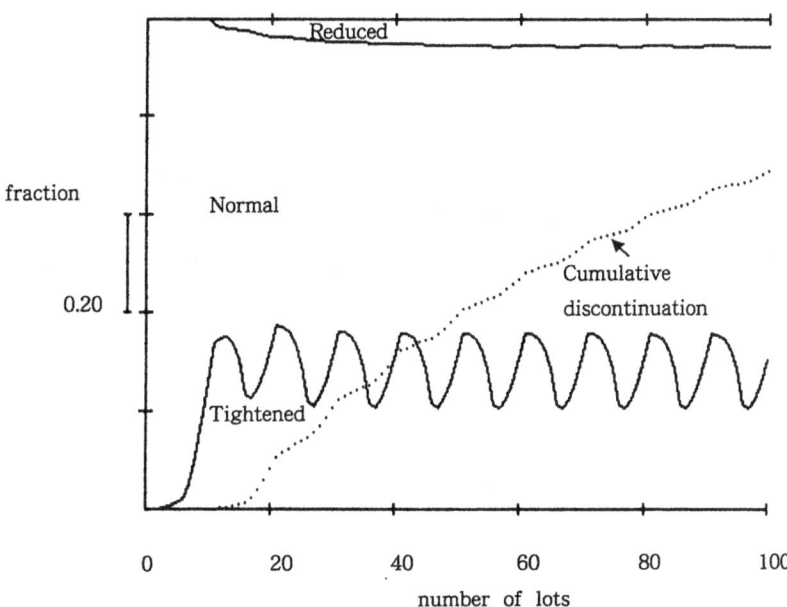

Figure 5: Sensitivity to the process changes every 5 lots ; $Ac_N = 2$, $Ac_T = 1$ and $Ac_R = 1$

48

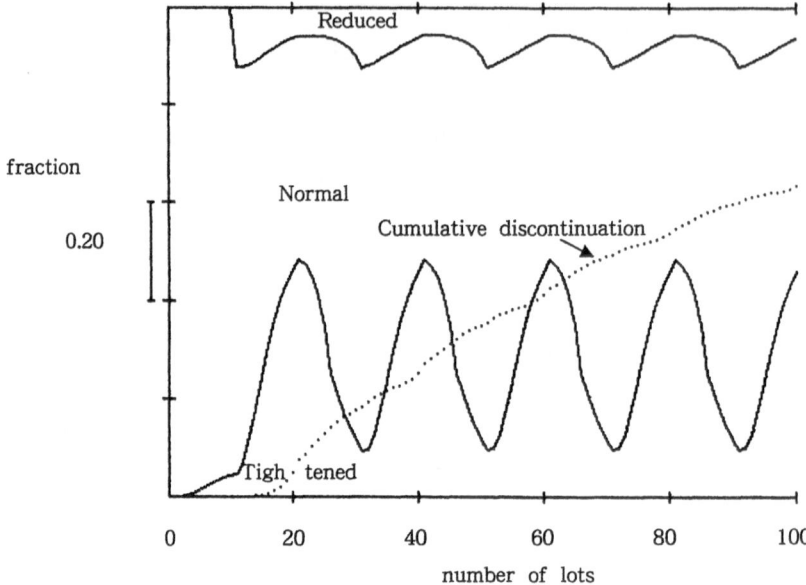

Figure 6: Sensitivity to process changes every 10 lots ; $Ac_N = 2$, $Ac_T = 1$ and $Ac_R = 1$

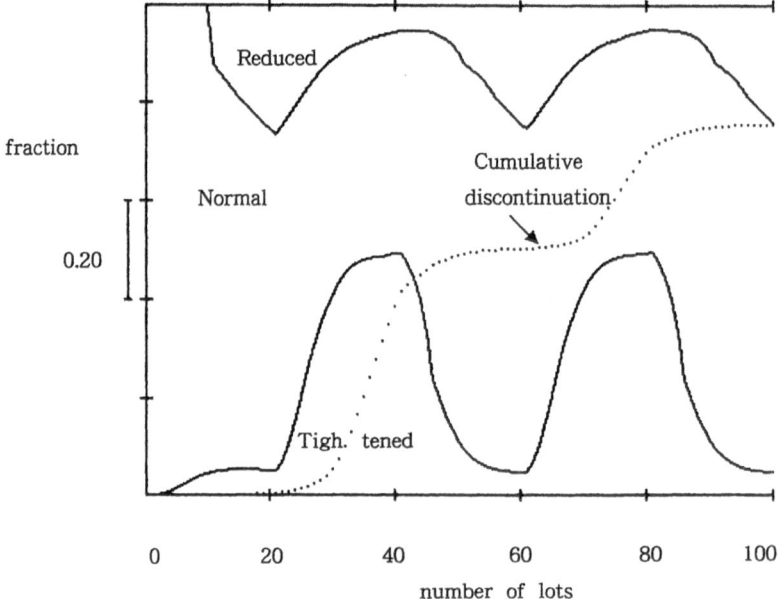

Figure 7: Sensitivity to process changes every 20 lots ; $Ac_N = 2$, $Ac_T = 1$ and $Ac_R = 1$

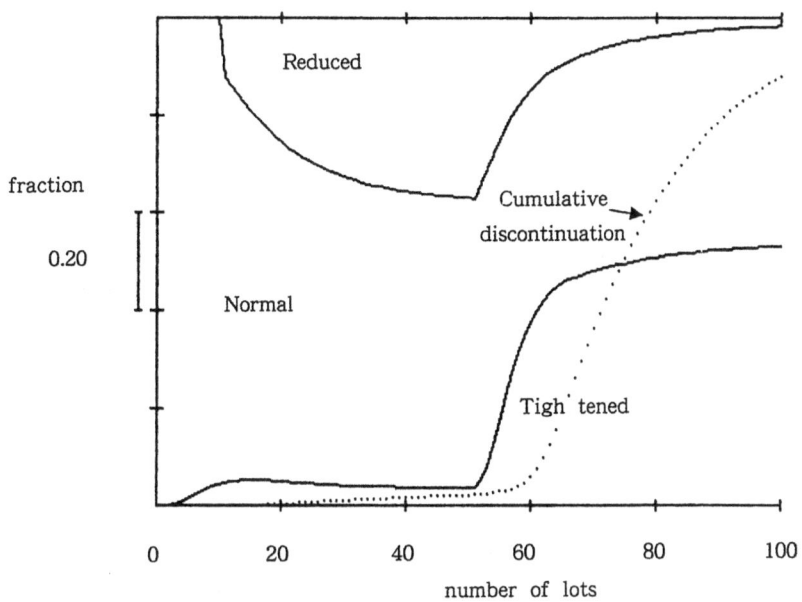

Figure 8: Sensitivity to process change at 50 − th lot of inspection ; $Ac_N = 2$, $Ac_T = 1$ and $Ac_R = 1$

50

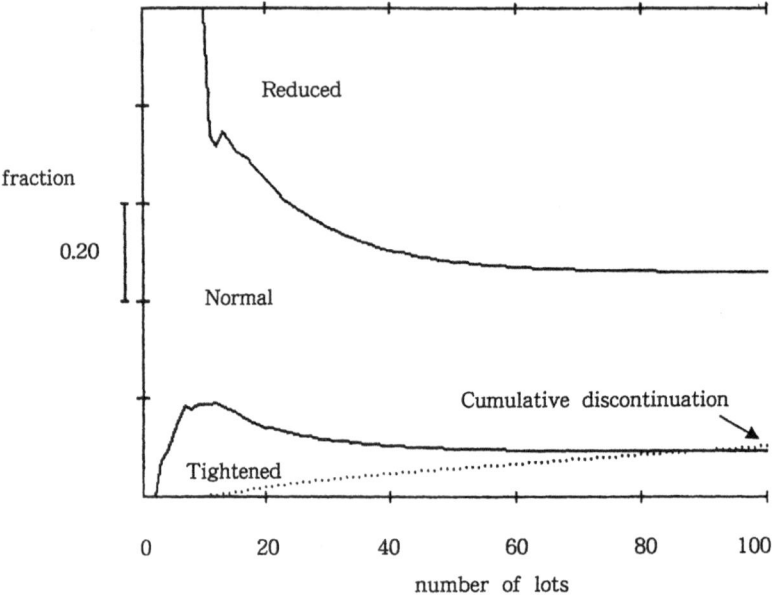

Figure 9: Fraction of inspection for p/AQL = 1.0 ; $Ac_N = 1/2$, $Ac_T = 1/3$ and $Ac_R = 1/3$

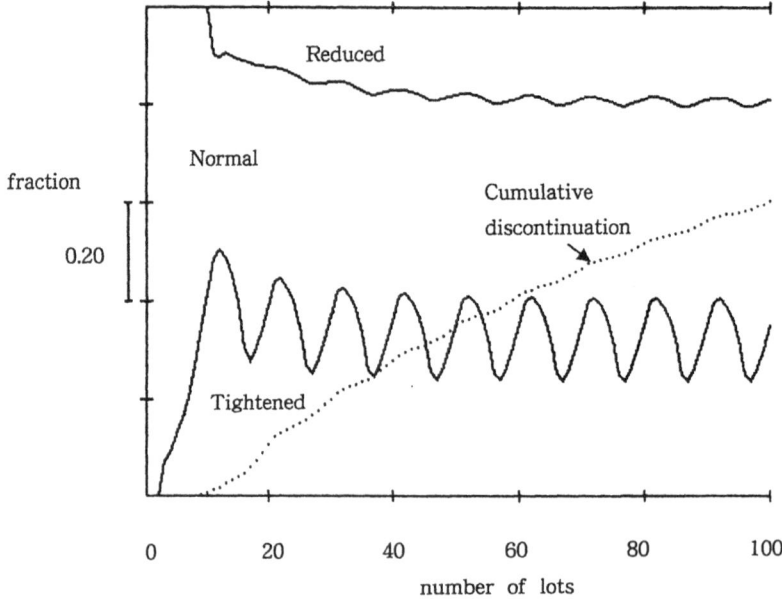

Figure 10: Sensitivity to the process changes every 5 lots ; $Ac_N = 1/2$

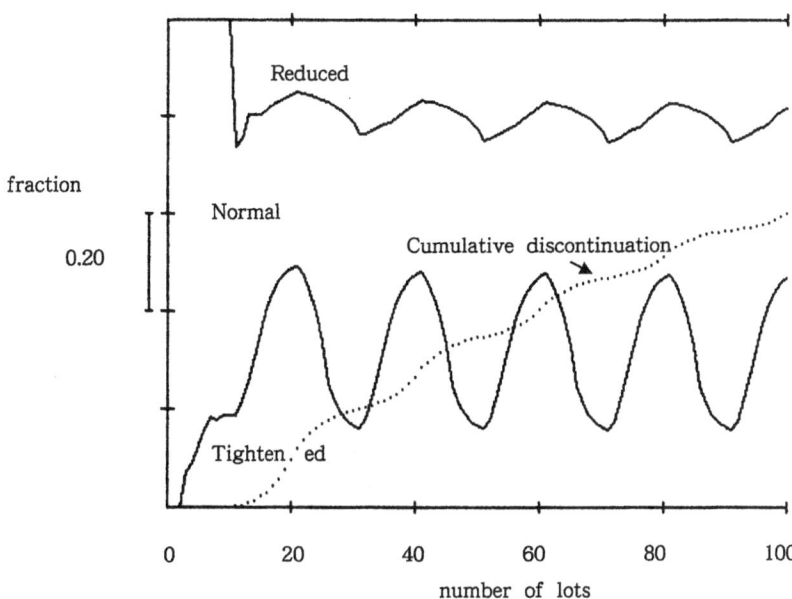

Figure 11: Sensitivity to process changes every 10 lots ; $Ac_N = 1/2$

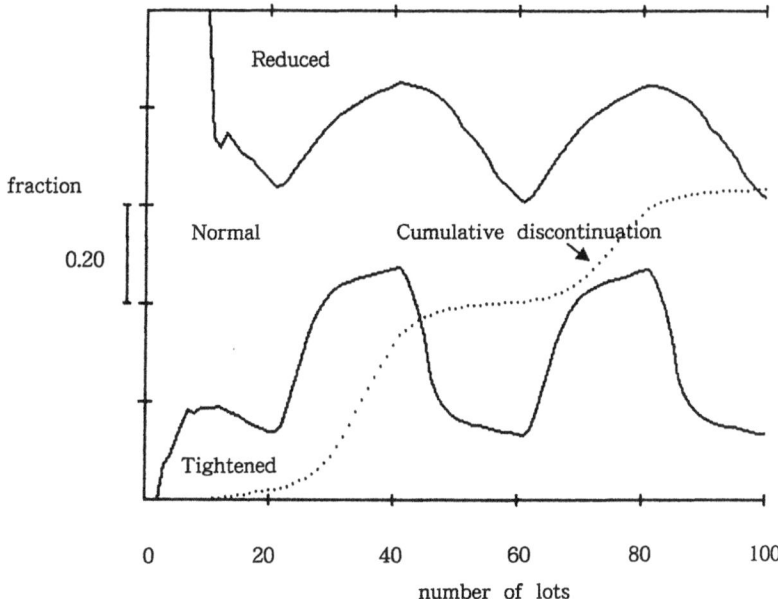

Figure 12: Sensitivity to process chanes every 20 lots ; $Ac_N = 1/2$

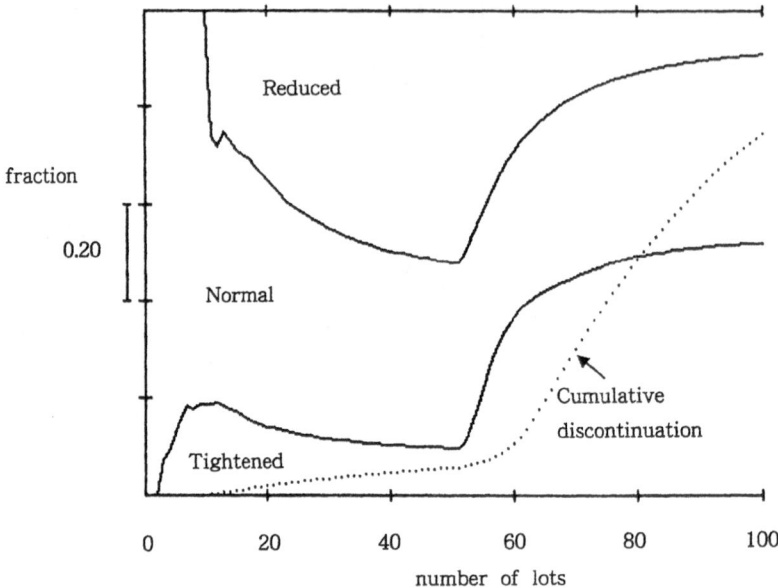

Figure 13: Sensitivity to process change at the $50-$th lot ; $Ac_N = 1/2$

tightened inspection increases. However, the switching schemes under the proposed revision of ISO 2859 − 1 guarantees higher sensitivity to deterioration in the process than those under ISO 2859, ISO 3951 or MIL − STD − 105D.

4.3 AAPs and ASSs

Figure 14 shows AAP, namely, the composite OC curves for $Ac_N = 2$, $Ac_T = 1$ and $Ac_R = 1$. Figure 15 presents ASS.

4.4 Comparison of sensitivity under the proposed revision and ISO 2859

The switching procedure from tightened to discontinuation in ISO 2859 or ISO 3951 is ; "In the event that 10 consecutive lots or batches remain on tightened inspection, inspection under the provision of this document should be discontinued ⋯". Koyama [12] gave the generating functions $G_{T-N}^*(x)$ and $G_{T-D}^*(x)$ under the switching rule, as follows ;

$$G_{T-N}^*(x) = P_T^5 x^5 + P_T^5 Q_T \sum_{\nu=6}^{10} x^\nu \ , \tag{56}$$

$$G_{T-D}^*(x) = (1 - 6P_T^5 + 5P_T^6) x^{10} \ . \tag{57}$$

Combined Equations (56) and (57) with Equations (26) and (33), we obtain

$$S_{T-N}^* = 6P_T^5 - 5P_T^6 \ , \tag{58}$$

$$S_{T-D}^* = 1 - 6P_T^5 + 5P_T^6 \ , \tag{59}$$

$$ARL_{T-N}^* = (45 - 40P_T) / (6 - 5P_T) \ , \tag{60}$$

$$ARL_{T-D}^* = 10 \ . \tag{61}$$

If we substitute the result of Equations (59), (60) and (61) into S_{T-D}, ARL_{T-D} and ARL_{T-N} in Table 6, we obtain ARL_{NRT-D}^* in Table 10, which is ARL from normal inspection to discontinuation of inspection under ISO2859, ISO 3951 or MIL − STD − 105D. As a result, we recognize $ARL_{NRT-D}^* < ARL_{NRT-D}$ for $p/AQL < 2$, while $ARL_{NRT-D}^* > ARL_{NRT-D}$ for $p/AQL > 2.5$, so that switching rule under the proposed revision of ISO 2859 − 1 assures higher sensitivity than that under ISO 2859, ISO 3951 or MIL − STD − 105D.

5. COMPUTATION TIME

Figure 16 shows that computation times to be required to calculate ARL_{NRT-D}s with a PC − 9805RA5 (CPU 80386 − 16MHz) personal computer, using BASIC language programs with double precision. Even if we use more capable computers, the relative value of computation times will be almost the same. A value of 10^{-7} is used to end computation, when ARL_{NRT-D}s were computed with transition probabilities. Computation times will grow much higher if we set the value lower. When the size of the determinant introduced in 3.3 is 34×34 for $Ac_N = 1/2$, 1817 seconds was consumed to compute S_{N-R}, S_{N-T}, ARL_{N-R} and ARL_{N-T}, while direct calculation necessitated 546 seconds for $p/AQL = 1.25$ and 4353 seconds for $p/AQL = 1.0$. Consequently, the direct calculation method is beneficial to us for $p/AQL > 1$. For $Ac_N \geqq 2$, the size of the determinant was reduced

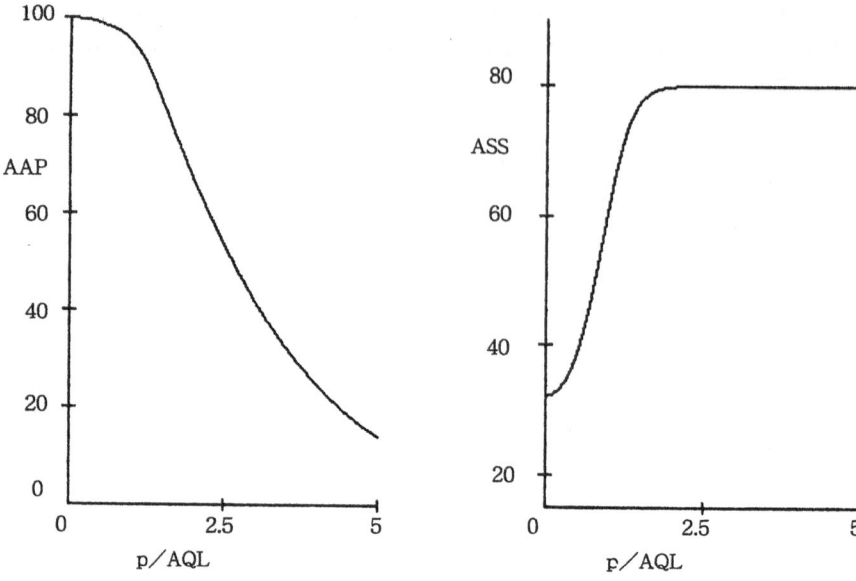

Figure 14: AAP for $Ac_N = 2$ Figure 15: ASS for $Ac_N = 2$

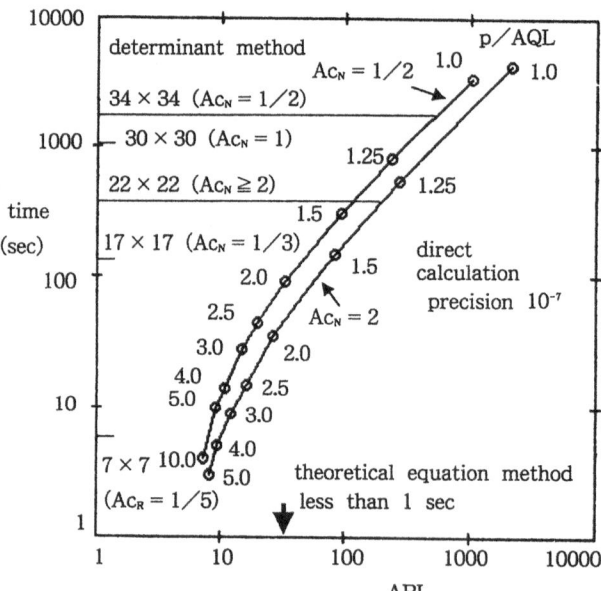

Figure 16: Computation times for calculating

to 22×22, computation time decreased about $1/5$ of that for $Ac_N = 1/2$.

If we can use the theoretical equations given by the generating function method, computation time to be required is less than 1 second. The value is much less than that by the two other methods, despite of complexity in calculating theoretical equations.

6. CLOSING REMARKS

The analyzed ESPs, AAPs, ASSs and ARLs with three different methods, namely, the direct calculation method with transition probabilities, the generating function method, and the determinant method. Although characteristics under the switching schemes in the existing specifications for sampling inspection are essentially transient, the calculated results in Figure 4 to 13 look rather ergodic where we considered the fractional probabilities only with $\phi_N(t)$, $\phi_T(t)$ and $\phi_R(t)$ except $\phi_D(t)$.

Higher sensitivity has been realized under the proposed revision of ISO 2859 compared with ISO 2859, ISO 3951 or MIL $-$ STD $-$ 105D, however, we should pay attention to process changes occurred in the period less than 20 lots of inspection.

The analyses presented in this paper will be useful to compare or match characteristics between the proposed revision of ISO 2859 $-$ 1 and ISO 3951.

7. APPENDIX : ABBREVIATIONS AND SYMBOLS

ARL : Average run length

AAP : Average acceptance probability

ASS : Average sample size

SFG : signal flow graph

ESP : Expected switching probability

AQL : Acceptable quality level

N : Normal inspection

T : Tightened inspection

R : Reduced inspection

D : Dicontinuation of inspection

Ac_N : Acceptance number on normal inspection

Ac_T : Acceptance number on tightened inspection

Ac_R : Acceptance number on reduced inspection

Ac_L : Acceptance number on normal inspection of one step tighter AQL

ARL_{i-j} : ARL from i to j : i, j = N, T, R or D

ARL_{T-j}^* : ARL from tightened inspection to j under 10 lots rule to discontinuation ; j = N, D

S_{i-j} : ESP from i to j ; i, j = N, T, R or D

S_{T-j}^* : ESP from tightened inspection to j under 10 lots rule to discontinuation ; j = N, D

$G_{i-j}(x)$: Generating function from i to j ; i, j = N, T, R or D

G_{T-j}^* : Generating function from tightened inspection to j under 10 lots rule to discontinuation ;

$i = N, D$

P_N : Probability of acceptsnce on normal inspection

P_T : Probability of acceptance on tightened inspection

P_R : Probability of acceptance on reduced inspection

P_L : Probability of acceptance on normal inspection of one step tighter AQL

P_M : $P_N - P_L$

Q_N : $1 - P_N$

Q_T : $1 - P_T$

Q_R : $1 - P_R$

n_N : Sample size on normal inspection

n_T : Sample size on tightened inspection

n_R : Sample size on reduced inspection

P_N (t) : Probability of acceptance on normal inspection at $t-$ th lot of inspection

P_T (t) : Probability of inspection on tightened inspection at $t-$ th lot of inspection

P_R (t) : Probability of inspection on reduced inspection at $t-$ th lot of inspection

P_L (t) : Probability of inspection on normal inspection of one step tighter AQL at $t-$ th lot of inspection

ϕ_N (t) : Probability on normal inspection at $t-$ th lot of inspection

ϕ_T (t) : Probability on tightened inspection at $t-$ th lot of inspection

ϕ_R (t) : Probability on reduced inspection at $t-$ th lot of inspection

ϕ_D (t) : Probability on discontinuation of inspection at $t-$ th lot of inspection

$Frac_N$ (t) : ϕ_N (t) $/$ [ϕ_N (t) $+ \phi_T$ (t) $+ \phi_R$ (t)]

$Frac_T$ (t) : ϕ_T (t) $/$ [ϕ_N (t) $+ \phi_T$ (t) $+ \phi_R$ (t)]

$Frac_R$ (t) : ϕ_R (t) $/$ [ϕ_N (t) $+ \phi_T$ (t) $+ \phi_R$ (t)]

AAP (t) : Average acceptance probability at $t-$ th lot of inspection

ASS (t) : Average sample size at the $t-$ th lot of inspection

ϕ (i, t) : Probability on node i at the $t-$ th lot of inspection

$\boldsymbol{p}^{(t)}$: $\boldsymbol{p}^{(t)} = [\phi \ (i, t)]$

$p_{i,j}$ (t) : Transition probability from node i to node j at the $t-$ th lot of inspection

$\boldsymbol{P}_t$: $\boldsymbol{P}_t = [p_{i,j} (t)]$

REFERENCES

(1) INTERNATIONAL ORGANIZATION FOR STANDARDIZATION (1974) : ISO 2859 Sampling Procedures for Inspection by Attributes.

(2) DODGE , H. F. (1955) : "Chain Sampling Inspection Plan", I. Q. C., Vol. XI, No. 4, p.10~13.

(3) MASON, S. J. (1956) : "Feedback Theory – Further Properties of Signal Flow Graphs", Proc. of IRE, July 1956, p. 920 – 926.

(4) Proposed Revision of ISO 2859 – 1 (1989) : Document ISO/TC69/SC5/WG2 – N87.

(5) INTERNATIONAL ORGANIZATION FOR STANDARDIZATION (1989) : ISO 2859 – 1 Sampling

Procedures for Inspection by Attributes.

(6) INTERNATIONAL ORGANIZATION FOR STANDARDIZATION (1981) : ISO 3951 Sampling Procedures and Charts for Inspection by Variables for Percent Defective.

(7) UNITED STATES DEPARTMENT OF DEFENSE (1963) : MIL − STD − 105D Sampling Procedures and Tables for Inspection by Attributes.

(8) JAPANESE STANDARDS ASSOCIATION (1971) : JIS Z 9015 Sampling Inspection Plans by Attributes with Severity Adjustment.

(9) FLAGLE, C. D., HUGGINS, W. H. and ROY, R. H. (1950) : Operations Research and Systems Engineering, The Johns Hopkins Press.

(10) THYREGOD, P. (1988) : Document ISO／TC69／SC5／WG2 − N76.

(11) KOYAMA, T. (1974) : Analysis on Dynamic Characteristics of Severity Control in Sampling Inspection Plans with Severity Adjustment, Ph. D. Thesis, Keio University, Tokyo.

(12) KOYAMA, T. (1979) : "Switching Characteristics under MIL − STD − 105D", Technometrics, Vol. 21, No. 1, p.9 − 19.

Bounded Outgoing Quality Sampling Plans when the Prior Distribution of Lot Defectives is not Binomial

E.L. Kaijage, J.E. Hall, and M.Z. Hassan, Chicago, Illinois, USA

1. INTRODUCTION

Attribute sampling inspection plans employing rectification, form a major part of statistical quality control.

This paper deals with sampling inspection plans by attributes calling for 100 percent inspection and rectification of rejected lots.

Dodge and Roming (1959) present an excellent treatise on inspection by attributes sampling plans employing rectification. They are:

a. Single Sampling Lot Tolerance Percent Defective ($LTPD$) plans

b. Double Sampling $LTPD$ plans

c. Single Sampling Average Outgoing Quality Limit ($AOQL$) plans

and

d. Double Sampling $AOQL$ plans.

The $LTPD$ sampling plans provide quality protection by insuring that a designated lot quality, will have a probability of acceptance not exceeding a given consumer risk, β.

The $AOQL$ is the maximum value of the average outgoing quality (AOQ) over all values of the process average. $AOQL$ sampling plans offer protection by guaranteeing that AOQ will be bounded above by a specified value equal to $AOQL$, irrespective of the value of the process average.

Dodge and Roming (1959) plans are constructed such that they give minimum average total inspection (ATI), per lot for products of a specified process average. $AOQL$ and $LTPD$ rectifying plans are well known. Duncan (1974) provides a comprehensive explanation of these plans. Their common denominator however is that they are all concerned with the expected value of the random variable, outgoing quality.

Hall (1979) points out that the outgoing quality is a random variable which has distributional properties other than the expected value . Attributes sampling plans employing rectification, use only the expected value of the outgoing quality and fail to account for lot-by-lot variability in outgoing quality.

In addition to using the expected value of the outgoing quality (AOQ), Hall (1979) uses the variance or standard deviation of outgoing quality, together with AOQ to develop "minimum variance" sampling plans. The standard deviation of the random variables, outgoing quality and total inspection, measure how close

Frontiers in Statistical Quality Control 4
Ed. by Lenz et al.
© Physica-Verlag Heidelberg 1992

the actual outgoing quality and the total inspection of a given lot are to the AOQ and ATI, respectively.

Hall and Hassan (1981) go a step further. They are concerned by the fact that, whereas $AOQL$ and $LTPD$ have great utility and application, when dealing with average properties of large streams of lots produced by a random process, these plans are ineffective as a measure of quality for a given lot or for a small group of lots. To improve conditions they propose designing single sampling inspection plans employing rectification which are based on the cumulative probability distribution of outgoing quality along with an outgoing quality upper bound. These sampling plans offer protection by guaranteeing that the outgoing quality of the lot (a random variable) will be bounded above by a specified value with some minimum probability, regardless of the value of the process average. This type of attributes sampling plan is known as a bounded outgoing quality sampling plan.

In developing the bounded outgoing quality plans Hall and Hassan (1981) assumed a binomial prior distribution of defectives in the lot. Mood (1943) states and proves the following theorem,

If X has the binomial distribution

$$f\{X\} = \binom{N}{X} p^X (1-p)^{N-X} \tag{1}$$

then x and $X - x$ are independently distributed,

where

$X =$ number of defectives in a lot and

$x =$ number of defectives in a sample.

That is, for a binomial distribution of lot defectives, samples from the lots will provide no basis whatsoever for inference about quality in the remainder of the lot (the unsampled part of the lot). This is a severe limitation. In this paper we will develop bounded outgoing quality plans when the process fraction defective p is not constant, that is when the prior distribution of defectives is not binomial. We assume that the binomial mass function defined in equation (1) is the conditional mass function of X given p,

$$f(X|p) = \binom{N}{X} p^X (1-p)^{N-X}, \quad X = 0, 1, \ldots N, \tag{1}$$

where p is a random variable with density funtion $f(p)$, defined on the unit interval $(0,1)$. The newly defined $f(X)$ now takes the form,

$$f(X) = \int_0^1 f(X|p)f(p)dp \tag{2}$$

and represents the non-binomial prior distribution of lot defectives. A further advantage of the procedure described by equation 2 is that any function established under the binomial prior assumption can be regarded as a conditional probability statement given p and will become an appropriate unconditional statement under integration, thus,

$$\phi(X) = \int_0^1 \phi(X|p)f(p)dp \qquad (3)$$

where $\phi(X|p)$ is an existing function of p derived under the binomial prior assumption and regarded as a conditional function of p. Specifically,

1. When the expected value of the process fraction defective is unknown, we demonstrate that the optimum sampling plans as derived by Hall and Hassan (1981) under the fixed p assumption are still optimum under the general assumption of an arbitrary prior distribution function on p.

2. When the process average $E(p)$ is known, and the beta-binomial prior distribution of lot defectives is assumed, that is $f(p)$ is defined as the beta density with shaping parameters a and b,

$$f(p) = \begin{cases} \dfrac{p^{a-1}(1-p)^{b-1}}{B(a,b)}, & 0 \le p \le 1 \\ 0, & \text{otherwise.} \end{cases} \qquad (4)$$

where $a, b > 0$, we show that optimum sampling plans which bound the outgoing quality with fixed probability, regardless of the beta density shaping parameters, can be determined.

2. DETERMINING SAMPLING PLANS WHEN THE PROCESS FRACTION DEFECTIVE p IS A RANDOM VARIABLE AND THE PROCESS AVERAGE $E(p)$ IS UNKNOWN

Assuming that the prior distribution of lot defectives is binomial, hence implying a constant process fraction defective p, Hall and Hassan (1981) devise single attributes plans with rectification based on the cumulative probability of distribution of outgoing quality (OQ) together with an outgoing quality upper bound.

Through hypergeometric sampling Hall (1979) derives the probability distribution of outgoing quality as,

$$P\{OQ = j/N\}$$
$$= \begin{cases} \sum_{x=0}^{c} \binom{n}{x} p^x (1-p)^{n-x} ((1-p)^{N-n} - 1) + 1, & j=0 \\ \sum_{x=0}^{c} \binom{n}{x} p^x (1-p)^{n-x} \binom{N-n}{j} p^j (1-p)^{N-n-j}, & j=1,...,\text{N-n} \end{cases} \qquad (5)$$

where,

     c = acceptance number

     j = number of defectives in the lot after sampling and rectification

n = sample size

N = lot size

OQ = outgoing quality

p = process average

Equation 5 implies the cumulative form of the probability distribution,

$$P\{j \leq j_o; c, p, n, N\}$$

$$= 1 - \sum_{x=0}^{c} \binom{n}{x} p^x (1-p)^{n-x} (1 - \sum_{j=0}^{j_o} \binom{N-n}{j} p^j (1-p)^{N-n-j}) \quad (6)$$

Hence they determine sampling plans that offer protection by guaranteeing that the outgoing quality for the lot is bounded above by a given (small) value with some minimum (large) probability irrespective of the value of the process average. Using the Poisson distribution to approximate the binomial prior distribution of defectives they present sampling plans (n, c) which satisfy,

$$\min_{p} P\{OQ \leq OQ_o\} \approx 1 - \alpha. \quad (7)$$

That is

$$\min_{p} P\{j \leq j_o\} \approx 1 - \alpha, \quad (8)$$

where

OQ_o = the desired upper bound to the outgoing quality OQ

j_o = upper bound to the number of defectives j in the lot after sampling and rectification

$1-\alpha$ = minimum value of $P\{OQ \leq OQ_o\}$.

In this paper we show that the sampling plans designed under the criteria of expression 7 assuming a binomial prior distribution of defectives in the lot, are invariant under the specific form of the prior distribution of defectives. That is the optimum sampling plans as derived through the principle of Hall and Hassan (1981) are optimum regardless of the type of prior distribution on the process fraction defective p.

Let the cumulative distribution of the outgoing quality (OQ) when the process fraction defective p is given be,

$$F_o(p) = P\{OQ \leq \frac{j_o}{N} | p\}. \quad (9)$$

A graph of the cumulative function is typified by Figure 1.

Let p^* be the value of p at which $F_o(p)$ is minimum. Then

$$\min_{p} F_o(p) = F_o(p^*).$$

It follows that,

$$E(F_0(p)) \geq \min_p F_0(p) = F_0(p^*).$$

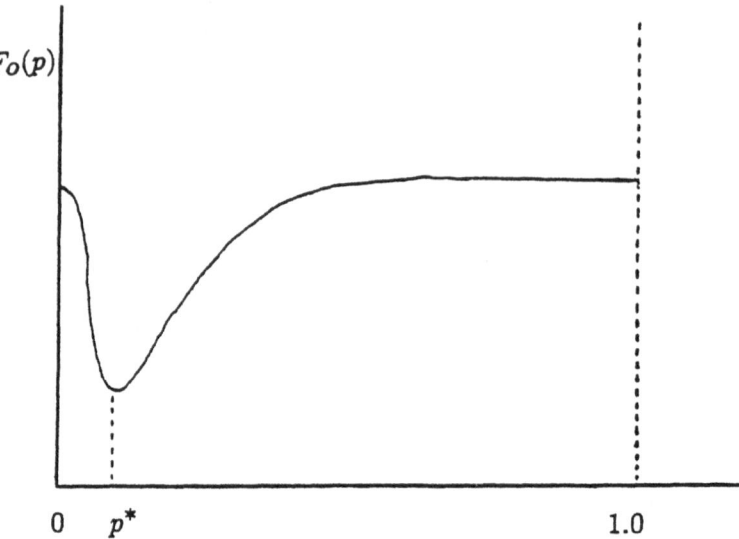

Figure 1: Relationship between $F_0(p)$ and p

Hence sampling plans derived through the principle of Hall and Hassan (1981) are optimum regardless of the type of prior distribution on the process fraction defective p.

In this paper we propose the following algorithm for determining the sampling plans,

Step 1. Let the prior distribution of lot fraction defectives be binomial. Choose the j_0 and $1 - \alpha$ values to be maintained.

Step 2. Let $c = 0$.

Step 3. Apply the method of golden section to search within the interval $0 \leq p \leq 1.0$, the value of p which yields the maximum sample size n_c such that,

$$P\{j \leq j_0\} = 1 - \sum_{x=0}^{c} \binom{n}{x} p^x (1-p)^{n-x} \left(1 - \sum_{j=0}^{j_0} \binom{N-n}{j} p^j (1-p)^{N-n-j}\right)$$
$$\tag{6}$$

$$\approx 1 - \alpha.$$

Also compute the corresponding average total inspection,

$$ATI_c = n_c + (N - n_c) \left(1 - \sum_{x=0}^{c} \binom{n_c}{x} p^x (1-p)^{n_c-x}\right). \tag{10}$$

Step 4. If $(ATI_{c-2} < ATI_{c-1} < ATI_c)$ or $(ATI_{c-1} \approx ATI_c)$ then let

$$c_* = c - 1 \quad \text{and} \quad n_* = n_{c-1}.$$

Proceed to step 5. Otherwise let $c = c + 1$ and proceed to step 3.

Step 5. The criteria of selection among the sampling plans (c, n_c) derived thus far, is minimum ATI_c. However if c is increasing and ATI_c is approximately equal to ATI_{c-1}, implying any larger values of c do not show a significantly smaller ATI_c, then select a sampling plan at ATI_{c-1}. Hence (c_*, n_*) is the required plan.

Table 1 shows examples of sampling plans (c, n_c) determined by this algorithm.

3. DETERMINING SAMPLING PLANS WHEN THE PROCESS FRACTION DEFECTIVE p IS A RANDOM VARIABLE AND THE PROCESS AVERAGE $E(p)$ IS KNOWN

When the process average $E(p)$ is known we assume a beta prior distribution of the process fraction defective. The beta density function is chosen because of its universal shaping parameters. Hence many varieties of distributions can be represented. The beta-binomial prior distribution of the number of defectives from lot to lot is defined by Hald (1960) as,

$$f_N(X) \doteq \binom{N}{X} \frac{B(x + a, N - X + b)}{B(a, b)}, \tag{11}$$

where,

$a =$ first shaping parameter of a beta distribution
$b =$ second shaping parameter of a beta distribution
$X =$ number of defectives in lot of size N.

He also defines the expected value and the variance of the process fraction defectives p as,

$$E(p) = \frac{a}{a + b} \tag{12}$$

and

$$\sigma_p^2 = \frac{ab}{(a + b)^2(a + b + 1)} \tag{13}$$

which implies

$$\sigma_p^2 = \frac{[E(p)]^2[1 - E(p)]}{E(p) + a}. \tag{14}$$

jo		c = 0	c = 1	c = 2	c = 3	c = 4	c = 5	(c_*, n_*)
1	n_c	628	770	836	874			
	ATI_c	909.03	911.88	922.93	934.83			(0,628)
	$P\{j \le jo\} \ge$	0.9503	0.9504	0.9506	0.9507			
2	n_c	484	637	721	774			
	ATI_c	905.82	899.71	904.05	906.12			(1,637)
	$P\{j \le jo\} \ge$	0.9503	0.9504	0.9505	0.9506			
3	n_c	394	541	630	691	735		
	ATI_c	901.22	891.39	889.61	895.59	899.79		(2,630)
	$P\{j \le jo\} \ge$	0.9504	0.9504	0.9505	0.9506	0.9506		
4	n_c	334	471	560	623	671		
	ATI_c	908.44	895.60	895.64	890.95	897.83		(3,623)
	$P\{j \le jo\} \ge$	0.9505	0.9505	0.9505	0.9506	0.9506		
5	n_c	290	417	503	567	616	656	
	$ATIc$	908.93	895.56	888.65	890.63	886.22	896.97	(4,616)
	$P\{j \le jo\} \ge$	0.9505	0.9505	0.9505	0.9506	0.9506	0.9507	

Table 1. Sampling Plans When the Process Average is Unknown and the Lot Size N = 1000

Through hypergeometric sampling Skellam (1948) derives the marginal distribution of the number of defectives from sample to sample as,

$$f_n(x) = \binom{n}{x} \frac{B(x + a, n - x + b)}{B(a, b)} \tag{15}$$

The probability distribution of outgoing quality (OQ) under these assumptions is,

$$P\{OQ = \frac{j}{N}\} = \begin{cases} (1 - P_A) + \sum_{x=0}^{c} \frac{B(x+a,n-x+b)}{B(a,b)}, & j = 0 \\ \binom{N-n}{j} \sum_{x=0}^{c} \binom{n}{x} \frac{B(j+x+a,N-j-x+b)}{B(a,b)}, & j = 1, 2, ..., N - n, \end{cases} \tag{16}$$

where,

$$P_A = \sum_{x=0}^{c} \binom{n}{x} \frac{B(x + a, n - x + b)}{B(a, b)}$$
$$= \text{Probability of lot acceptance.}$$

The probability that the outgoing quality (OQ) is equal to or less than a given level of quality (OQ_o) is,

$$P\{OQ \leq OQ_o; a, b, c, n, N\}$$
$$= P\{j \leq j_o; a, b, c, n, N\}$$
$$= 1 - \sum_{x=0}^{c} \binom{n}{x} \frac{B(x+a, n-x+b)}{B(a,b)}$$
$$+ \sum_{j=0}^{j_o} \binom{N-n}{j} \sum_{x=0}^{c} \binom{n}{x} \frac{B(j+x+a, N-j-x+b)}{B(a,b)}. \quad (17)$$

Lauer (1978) discusses acceptance probabilities for sampling plans where the process fraction defective p is a random variable which varies from lot to lot following a beta distribution. He derives the range of the variance of p as,

$$0 \leq \sigma_p^2 \leq E(p)(1 - E(p)). \quad (18)$$

From equations (12) and (13) it follows that

$$a = E(p) \left[\frac{(1 - E(p))E(p)}{\sigma_p^2} - 1 \right] \quad (19)$$

and

$$b = a \frac{1 - E(p)}{E(p)} \quad (20)$$

We now employ the following algorithm to determine the optimum sampling plan under the assumption that the process average $E(p)$ is known.

Step 1. The lot size N is known. Initially assume the variance of the process fraction defective $\sigma_p^2 = 0$. Hence assume a binomial prior distribution of the number of defectives from lot to lot. Choose j_o and $1 - \alpha$ values desired.

Step 2. Let $c = 0$.

Step 3. Starting from sample size $n = 1$, compute n_c such that

$$P\{j \leq j_o | E(p) = \text{ known }\}$$
$$= 1 - \sum_{x=0}^{c} \binom{n}{x} (E(p))^x (1 - E(p))^{n-x} \{1$$
$$- \sum_{j=0}^{j_o} \binom{N-n}{j} (E(p))^j (1 - E(p))^{N-n-j}\}$$
$$\approx 1 - \alpha. \quad (6)$$

Let the sample size n which satisfies equation 6 when the acceptance number is c be n_C. Also for sample size n_C compute the average total inspection,

$$ATI = n + (N - n)(1 - \sum_{x=0}^{c} \binom{n}{x} p^x (1 - p)^{n-x}).\tag{10}$$

Step 4. Now, assume a beta-binomial prior distribution of the number of defectives from lot to lot. That is assume

$$P\{j \leq jo|c,\ n,\ N\}$$

$$= 1 - \frac{1}{B(a, b)} \{ \sum_{x=0}^{c} \binom{n}{x} B(x + a, n - x + b)$$

$$- \sum_{j=0}^{jo} \binom{N-n}{j} \sum_{x=0}^{c} \binom{n}{x} B(j + x + a, N - j - x + b)\}.\tag{17}$$

Let

$$a = E(p) \left(\frac{E(p)(1 - E(p))}{\sigma_p^2} - 1 \right),\tag{19}$$

and

$$b = a \left(\frac{1 - E(p)}{E(p)} \right)\tag{20}$$

where σ_p^2 lies within the following range

$$0 < \sigma_p^2 < E(p)(1 - E(p)).\tag{18}$$

Also for sample n_C compute the average total inspection,

$$ATI_C = n_C + (N - n) \left(1 - \sum_{x=0}^{c} \binom{n_C}{x} \frac{B(x + a, n_C - x + b)}{B(a, b)} \right)$$

Now, using the sample size n_C obtained from the preceding step, perform the method of golden section search to find the σ_p^2 at which minimum $P\{j \leq jo\}$ occurs.

If minimum $P\{j \leq jo|c,\ n_C,\ N\} < 1 - \alpha$, proceed to step 5.
If minimum $P\{j \leq jo|c,\ n_C,\ N\} \geq 1 - \alpha$, proceed to step 6.

Step 5. At the value of σ_p^2 where the minimum $P\{j \leq jo|c,\ n_C,\ N\}$ occurs in step 4, increase the sample size to the minimum sample size n_1 such that

$$P\{j \leq jo|\sigma_p^2,\ c,\ N\} \geq 1 - \alpha.$$

Let $n_c = n_1$ and proceed to step 4.

jo		c = 0	c = 1	c = 2	c = 3	$(c_\star, n_\star)$
1	n_c	358	544	656	730	
	ATL_c	**856.02**	882.32	906.73	926.51	(0,358)
	$P\{j \leq jo\} \geq$	0.9500	0.9502	0.9502	0.9503	
2	n_c	277	436	544		
	ATL_c	**868.38**	876.33	901.10		(0,277)
	$P\{j \leq jo\} \geq$	0.9503	0.9503	0.9501		
3	n_c	233	372	473	551	
	ATL_c	883.19	**881.63**	903.23	909.86	(1,372)
	$P\{j \leq jo\} \geq$	0.9502	0.9503	0.9503	0.9502	
4	n_c	206	330	424		
	ATL_c	898.17	**891.22**	896.72		(1,330)
	$P\{j \leq jo\} \geq$	0.9503	0.9504	0.9503		
5	n_c	187	300	388		
	$ATIc$	906.26	**896.72**	899.93		(1,300)
	$P\{j \leq jo\} \geq$	0.9501	0.9504	0.9504		
6	n_c	174	278	361		
	ATL_c	915.00	**904.15**	914.69		(1,278)
	$P\{j \leq jo\} \geq$	0.9502	0.9505	0.9505		
7	n_c	165	261	340	406	
	ATL_c	932.86	**922.75**	923.13	927.25	(1,261)
	$P\{j \leq jo\} \geq$	0.9506	0.9501	0.9505	0.9503	
8	n_c	157	249	**323**	388	
	ATL_c	940.55	936.43	**934.70**	938.60	(2,323)
	$P\{j \leq jo\} \geq$	0.9500	0.9506	0.9500	0.9506	
9	n_c	152	239	311	373	
	ATL_c	946.14	**943.63**	944.40	947.35	(1,239)
	$P\{j \leq jo\} \geq$	0.9505	0.9504	0.9505	0.9502	
10	n_c	148	232	301	362	
	ATL_c	949.94	**949.40**	950.78	954.48	(1,232)
	$P\{j \leq jo\} \geq$	0.9506	0.9508	0.9503	0.9502	

Table 2. Sampling Plans When the Process Average E(p) = 0.02, the Lot Size N = 1000 and the Prior Distribution is Beta-Binomial

Step 6. The sample plan (c, n_c) is optimum regardless of the values of the shaping parameters a and b. That is regardless of the value of σ_p^2.

If $(ATI_{c-2} < ATI_{c-1} < ATI_c)$ or $(ATI_{c-1} \approx ATI_c)$ then let,

$$c_* = c - 1, \quad n_* = n_{c-1}$$

and proceed to step 7. Otherwise let,

$$c = c + 1$$

and proceed to step 3.

Step 7. The required sampling plan is (c_*, n_*). The criterion for selection is minimum ATI_c.

Table 2 shows examples of sampling plans (c, n_c) determined by this algorithm.

4. APPENDIX

Artin (1964) derives,

$$B(x, y) = \frac{\Gamma(x)\Gamma(y)}{\Gamma(x + y)}. \tag{21}$$

U.S. Department of Commerce (1972) gives

$$\frac{1}{\Gamma(z)} = \sum_{k=1}^{\infty} C_k z^k; \ |z| < \infty. \tag{22}$$

From equation 22

$$\Gamma(z) = \frac{1}{\sum_{k=1}^{\infty} C_k z^k}.$$

Also the U.S. Department of Commerce (1972) gives C_k values as shown in Table 3. Hence from Table 3,

$$\Gamma(z) = \frac{1}{\sum_{k=1}^{26} C_k z^k}. \tag{23}$$

Assuming $x \leq y$ and computing the constants t_1, t_2, t_3 and the variables v_1, v_2, v_3 as,

$$
\begin{aligned}
t_1 &= <x> & v_1 &= x - t_1 \\
t_2 &= <y> & v_2 &= y - t_2 \\
t_3 &= <x + y> & v_3 &= x + y - t_3
\end{aligned}
\tag{24}
$$

where $< \ >$ is the greatest integer less than or equal to a variable defined within $< \ >$, Pachner (1984) determines the beta function as,

a. If $t_1 < t_2$, then

$$B(x,y) = \frac{\Gamma(v_1)\Gamma(v_2)}{\Gamma(v_3)} \cdot \frac{\left(\prod_{k=1}^{t_1} \frac{(x-k)(y-k)}{(x+y-k)}\right)\left(\prod_{k=t_1+1}^{t_2} \frac{y-k}{x+y-k}\right)}{\prod_{k=t_1+1}^{t_3}(x+y-k)}. \qquad (25)$$

b. If $t_1 = t_2$, then

$$B(x,y) = \frac{\Gamma(v_1)\Gamma(v_2)}{\Gamma(v_3)} \cdot \frac{\prod_{k=1}^{t_1} \frac{(x-k)(y-k)}{x+y-k}}{\prod_{k=t_1+1}^{t_3}(x+y-k)}. \qquad (26)$$

k	C_k
1	1.00000 00000 000000
2	0.57721 56649 015329
3	-0.65587 80715 202538
4	-0.04200 26350 340952
5	0.16653 86113 822915
6	-0.04219 77345 555443
7	-0.00962 19715 278770
8	0.00721 89432 466630
9	-0.00116 51675 918591
10	-0.00021 52416 761149
11	0.00012 80502 823882
12	-0.00002 01348 547807
13	-0.00000 12504 934821
14	0.00000 11330 272320
15	-0.00000 02056 338417
16	0.00000 00061 160950
17	0.00000 00050 020075
18	-0.00000 00011 812746
19	0.00000 00001 043427
20	0.00000 00000 077823
21	-0.00000 00000 036968
22	0.00000 00000 005100
23	-0.00000 00000 000206
24	-0.00000 00000 000054
25	0.00000 00000 000014
26	0.00000 00000 000001

Table 3. Values of C_k

c. If x and y are positive integers t and u, then

$$B(t, u) = \frac{1}{u\binom{t+u-1}{u}}. \tag{27}$$

The binomial coefficient $\binom{t+u-1}{u}$ is computed by equation 28. To prevent an overflow of the computer when $n > 33$, Pachner (1984) evaluates a binomial coefficient thus,

$$\binom{n}{k} = \begin{cases} \prod_{j=1}^{n-k} \frac{k+j}{j}, & \text{for } k > \frac{n}{2} \\[2mm] \prod_{j=0}^{k} \frac{n-k+j}{j}, & \text{for } k \leq \frac{n}{2}. \end{cases} \tag{28}$$

Equation 23 is used to compute $\Gamma(v_1)$, $\Gamma(v_2)$, $\Gamma(v_3)$ in equations 25 and 26.

Equation 28 is used to compute binomial coefficients and equations 24, 25, 26 and 27 are used to compute beta functions.

The golden section method is a procedure for minimizing strictly quasiconvex functions of one variable without using derivatives. Bazaraa and Shetty (1979) present an algorithm for the method.

5. REFERENCES

[1]   ARTIN, E. (1964): *The Gamma Function*, New York, NY: Holt, Rhinehart and Winston

[2]   BAZARAA, M. S. and SHETTY, C. M. (1979): *Nonlinear Programming Theory and Algorithms*, New York: John Wiley and Sons.

[3]   DODGE, H. F. and ROMING, H. G. (1959): *Sampling Inspection Tables*, New York: John Wiley.

[4]   DUNCAN, A. J. (1974): *Quality Control and Industrial Statistics*, Homewood, Illinois: Richard D. Irwin, Inc.

[5]   HALD, A. (1960), "The Compound Hypergeometric Distribution and a System of Single Sampling Inspection Plans Based on Prior Distributions and Costs," *Technometrics*, 2, 275-340.

[6]   HALL, J. E. (1979): "Minimum Variance and VOQL Sampling Plans," *Technometrics*, 21, 555-565.

[7]   HALL, J. E. and HASSAN, M. Z. (1981): "On the Cumulative Distribution of Outgoing Quality: A New Criterion for Sampling Plans," *Technometrics*, 23, 395-400.

[8]   LAUER, G. N. (1978): "Acceptance Probabilities for Sampling Plans Where the Proportion Defective Has a Beta Distribution," *Journal of Quality Technology*, 10, 52-55.

[9]   MOOD, A. M. (1943): "On the Dependence of Sampling Inspection
        Plans Upon Population Distributions," *Annals of Mathematical
        Statistics*, 14, 415-425.

[10]   PACHNER, J. (1984): *Handbook of Numerical Analysis Applications*,
        New York: McGraw-Hill Book Company.

[11]   SKELLAM, J. G. (1948): "A Probability Distribution Derived from
        the Binomial Distribution by Regarding the Probability of Success
        as a Variable between Sets of Trials," *Journal of the Royal
        Statistical Society*, Series B,10, 257-261.

[12]   National Bureau of Statistics, U.S. Department of Commerce (1972):
        *A Handbook of Mathematical Functions with Formulas,
        Graphs, and Mathematical Tables*, Washington, D.C:
        U.S Government Printing Office.

Developmental Studies on New Sampling Inspection Plans

H. Kimura, Okayama, Japan
Ch. Asano, Tokyo, Japan

1. INTRODUCTION

In view of industrial productivity and marketing flow of products in practice, sampling inspection plans (s.i.p.s) have carried out effectively their mission at every stage of acceptance, manufacturing processes and shipment of products. Accordingly, it is also wellknown that several plans by attributes and by variables are ready-made and proposed as international and domestic standards of industry, e.g. JIS Z-9003, MIL-STD-105D, ISO 3951, JIS Z-9015, although there exist delicate differences on nuance among Acceptance (Vendor and Vendee) s.i.p.s, Intermediate s.i.p.s, Final (Outgoing) s.i.p.s, and others (e.g. Detail inspection, Strange inspection, Half-finished product inspection, Lot inspection), and that such differences are extremely influential to apply scientific s.i.p.s to practice.

Nowadays, because of recent changes of manufacturing conditions and circumstances with high technics, different situations have appeared. In view of current existence of different states of process control, based on high reliability, computer-controlled high speed and mass production, the distribution laws of qualities are greatly improved. From this viewpoint, many products have been improved on their life-time and reliability, and some distribution types of quality may be considered adequately to be Weibull, exponential and so on, but not always normal, expecting approximate normality of sample means.

Accordingly, the development of more suitable s.i.p.s is required theoretically, and actually it seems to us that still there exist much room to develop inspection plans, especially for Lot-by-lot inspections, Continuous production inspections, Control inspections, e.g. Process inspections, Patrol sampling inspections, Spot check inspections, First piece inspections. Since s.i.p.s by variables and by attributes have been recently proposed on simple assumptions of distribution laws and test statistics, the mathematical formulations are very complicated. In the subsequent sections, some new s.i.p.s are illustrated by nonparametric and complex comparative studies for several distribution types in the cases of sensory or biometric life-test, omitting their mathematical formulations through their uselessness on numerical evaluation.

Thus the purpose of the present paper is firstly to introduce the software system Micro-NISAN on a personal computer as an effective tool with statistical knowledge functions to study statistical properties of s.i.p.s and to develop new plans with wide variety. Secondly, we illustrate and propose concretely some new s.i.p.s with non-parametric tests for several types of distribution. Thirdly adaptive s.i.p.s are investigated, including ISO 3951 and so on, because inferential properties of such ready-made s.i.p.s still seem to remain vague.

2. MICRO–NISAN SYSTEM AND THE CAPACITY

Micro-NISAN system is a general purpose interactive statistical analysis system implemented on micro computers. Regarding studies on s.i.p.s, the functions of statistical knowledge acquisition in Micro-NISAN are especially useful with high flexibility. Brief explanation about them are given in this section.

Frontiers in Statistical Quality Control 4
Ed. by Lenz et al.
© Physica-Verlag Heidelberg 1992

In order to define distribution types of quality within or between lots, a command **DEFRAN** defines a type of single distribution types, e.g. exponential, binomial, Poisson, uniform, Weibull and normal distributions. Practically, the definition of a lot is important, and actually there exist several kinds of lot, e.g. main material lots, manufacturing lots, outgoing lots and sales lots, where the last two kinds of lots are considered to be mixed, composite or complex. For this, a command **MIXRAN** defines their mixed distributions.

In either situation of acceptance, outgoing or process control, s.i.p.s are applied to practice, in view of power of test and required cost. A command **POWER** provides the power of testing hypothesis with underlying types of distribution. Two commands **REF** and **AREF** are useful to choose one of two arbitrary tests on the basis of relative efficiency and asymptotic relative efficiency, respectively.

In order to study a test process, a command **DEFTP** is to define a test process or a multi-stage test procedure, and gives a new name to use it in other commands, e.g. **POWER**, like a single test. A command **DIST** gives the distribution of test statistics and a command **CONDIST** is to show a conditional distribution of the final test statistic in a testing process defined by the **DEFTP**, where the conditional distribution gives a nominal probability and the individual risk probabilities for preceding tests in a test process.

For studies on sequential or multi-stage s.i.p.s, a command **PRTP** is to give power, average sample number and expected cost, including both costs of sample set-up and samples, including sampling rules of one-by-one-, partial- or group-sampling and with both decision rules of Type I and Type II, under any distribution type and parameter values, initial and additional sample sizes, hypothesis and so on.

Regarding parameter estimation after a preliminary test process, **PRTE** command is to give expected values of parameters, the mean square error and value of risk function for estimation. **RDIST** is to show the distribution of the final test statistic at a test process like **DIST** for a multi-stage test process with Types I and II.

3. SINGLE SAMPLING INSPECTION PLANS

Sensory tests of time-deterioration or life-time are frequently applied for comparative study between two sample groups of standard and test products, and ordinarily single s.i.p.s are applied at a time of cross section, because of the simplicity. In order to show the flexible functions of the system for s.i.p.s, Wilcoxon test, t-test and Kolmogorov-Smirnov test have been investigated in practical situations, where the t-test is based on the approximate normality of sample means.

3.1 STUDY ON WEIBULL DISTRIBUTIONS

Assuming common and known values of shape and scale parameters of the Weibull distribution, the powers of Wilcoxon test and t-test are examined on different locations of the distributions by the use of a command POWER. The results are shown in Figure 1, and it will be seen that the Wilcoxon test is relatively superior than the t-test. As an examination, since the Weibull distribution has a wide variety on the shape, basing on the meaning or model of the phenomenon, the similar examinations for location test are tried on a different shape of two populations. Figure 2 shows the power of both tests. As expectedly, the result is quite similar.

3.2 STUDIES ON EXPONENTIAL DISTRIBUTION AND UNIFORM DISTRIBUTION

The similar studies on the power of location tests are tried for the uniform distribution and the mixed exponential distribution, in Figures 3 and 4, respectively. At the first glance, it will be seen originally that Wilcoxon test is superior than t-test for such distributions.

74

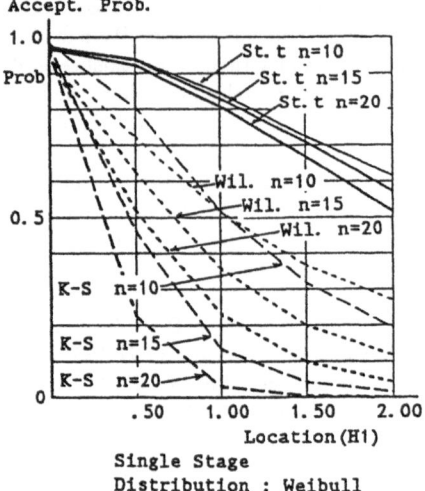

Figure 1: OC curves for Weibull location,
in case of shape = 0.5

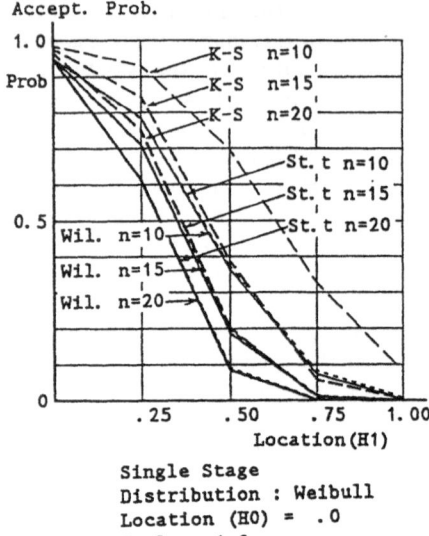

Figure 2: OC curves for Weibull location,
in case of shape = 2.0

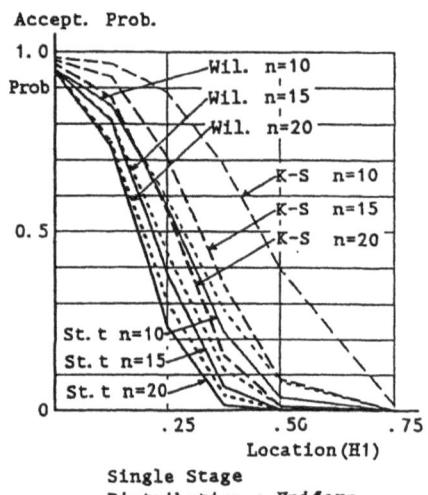

Figure 3: OC curves for uniform location

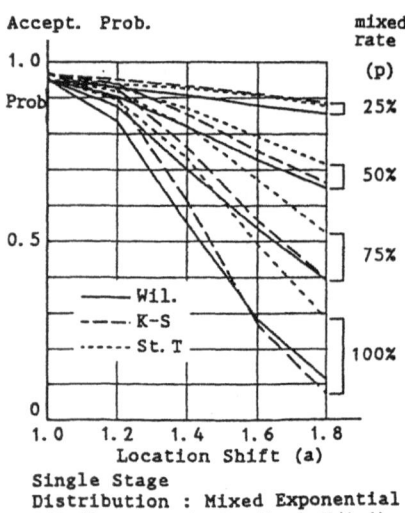

Figure 4: OC curves for mixed exponential
location

4. MULTI-STAGE SAMPLING INSPECTION PLANS

Multi-stage tests are practically attractive and effective in view of ASN and expected cost. Furthermore, persons in charge of this kinds of s.i.p.s are able to have consciousness of quality in practice, although single s.i.p.s are sometimes business-likely applied in routine work. Figures 5 and 6 show the OC curves and ASN curves of two-stage and five-stage s.i.p.s by Wilcoxon test, t-test, and Kolmogorov-Smirnov test for several Weibull distributions, like in the previous section. From these Figures, several insights will be obtained intuitively to design a s.i.p., although a randomized test or a command CONDIST is needed to give such a nominal probability of test statistic to consist both powers under null hypothesis. Such kinds of trials are easily and very quickly examined by using the Micro-NISAN system.

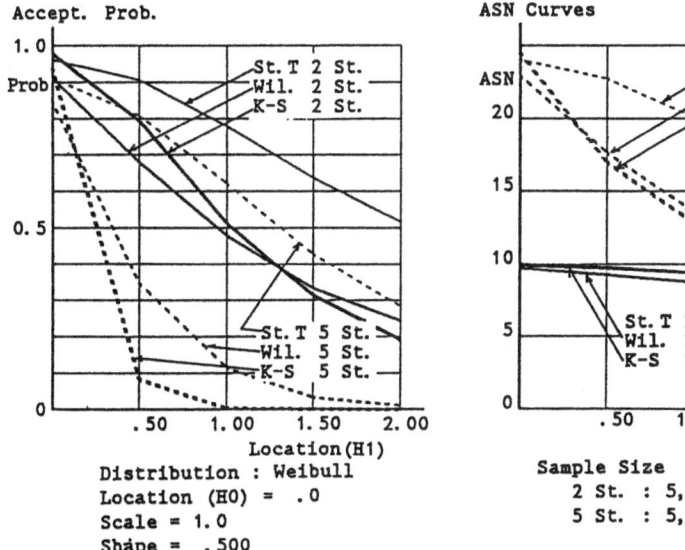

Figure 5: OC curves of two- and five-stage tests for Weibull location

Figure 6: ASN curves for the case in Fig. 5

5. ADAPTIVE SAMPLING INSPECTION PLANS

The principle of adaptive s.i.p.s, including continuous s.i.p.s, is ordinarily applied for a long-term trade, and several ready-made plans are provided, e.g. ISO 3951, MIL-STD-105D, JIS Z-9011, JIS Z-9015, under a normality of supplier's quality. These pragmatized plans are concerned with average level of quality after a certain screening test. The procedures are adjustable for the severity of inspection and the rules are rather complex with several operational and dynamic procedures, depending on variation of supplier's quality.

Ready-made plans are typically simplified on both sizes of lot and samples and limited to an assumption on the distribution laws of quality. Regarding such s.i.p.s, however, it seems to us that there exist unclear criteria on statistical optimality for the definite rules, e.g. (i) how to give the relations among statistical severities of normal inspection (NI), reduced inspection (RI) and tightened inspection (TI) plans, and (ii) how to determine the optimal timing of transitions between the inspection procedures.

In this section, some properties of ISO 3951 are investigated, and then the improved plans are proposed with an adequate criterion.

5.1 PROPERTIES OF ISO 3951 FOR NORMAL DISTRIBUTIONS

Several plans of ISO 3951 are studied on the normal, reduced, and tightened procedures for several lot sizes and AQLs. As an illustration, Figure 7-(a) shows three OC curves in case of N = 500, the inspection level = III, sample size code = J, AQL = 4.0%. At the first glance of the figure, however, non-statistical foremen and workers will feel simply something funny because the sample size of the tightened inspection is smaller than that of the normal inspection, excepting the larger k. It is also pointed out that there exist a certain difference on the locations and slopes of three curves of a plan, although the curves are in order of tightened, normal and reduced with a constant AOQ required. From this view-point, in stead of such a plan, another plan may be considered as shown in Figure 7-(b), where the sample size of TI is increased and k is constant for NI, TI and RI. It will be seen that three OC curves coincide approximately. Moreover, the sample size of RI can be reduced, e.g. Figure 7-(c), if a prior distribution of quality is considered for such a reliable situation.

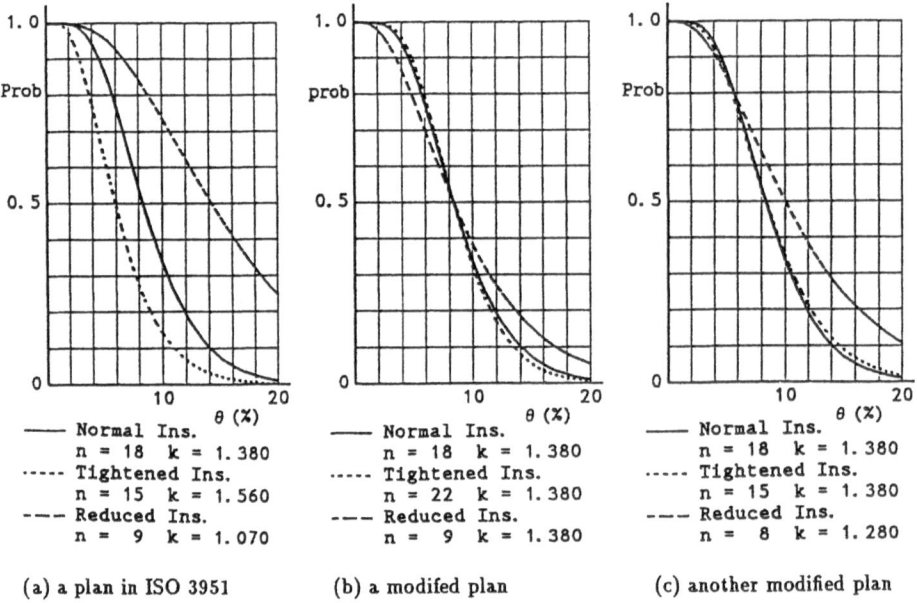

(a) a plan in ISO 3951   (b) a modifed plan   (c) another modified plan

Figure 7: OC curves for plans based on ISO 3951: Lot Size = 500, Inspection level= III, Sample Size Code=J, AQL= 4%

Depending on the inspection levels I, II and III, Figures 8-(a), 8-(b) and 8-(c) show OC curves of another plan, investigated by Koyama[4]. All of OC curves have the same relations among NI, TI and RI's, discussed above, cf., Figure 7-(a). Therefore, the improvement like Figure 7-(b) or 7-(c) may be commonly considered through the plans in ISO 3951.

5.2 TRANSITION PROBABILITIES

The ratios of transitions among normal, reduced and tightened inspections, and their dynamic distributions of ratios are investigated, as shown in Figure 9, where the OC curves of NI, RI and TI are shown in Figure 8-(b). The four names P(n), P(r), P(d) and P(r+d) of curves in the top graphs

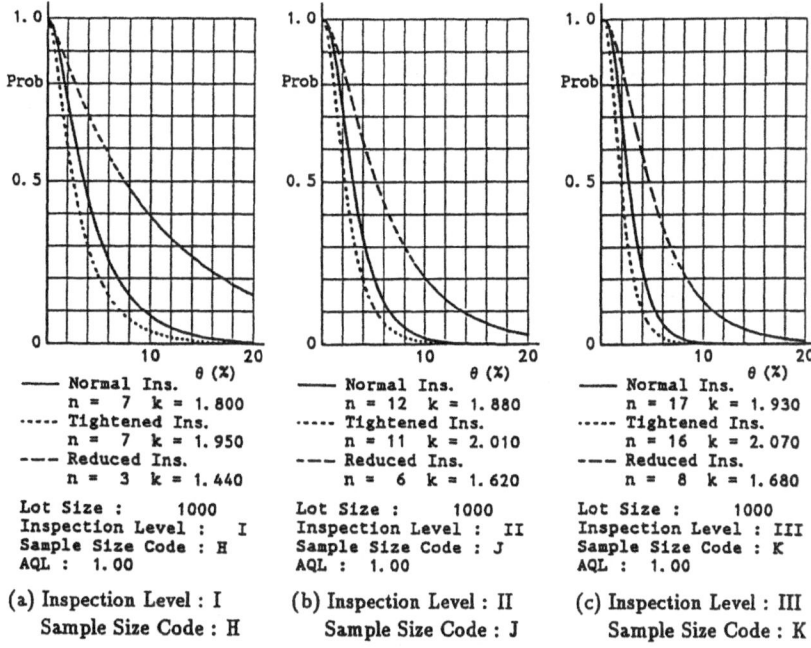

(a) Inspection Level : I
Sample Size Code : H

(b) Inspection Level : II
Sample Size Code : J

(c) Inspection Level : III
Sample Size Code : K

Figure 8: OC curves of plans based on ISO 3951: Lot Size = 1000, AQL= 1%

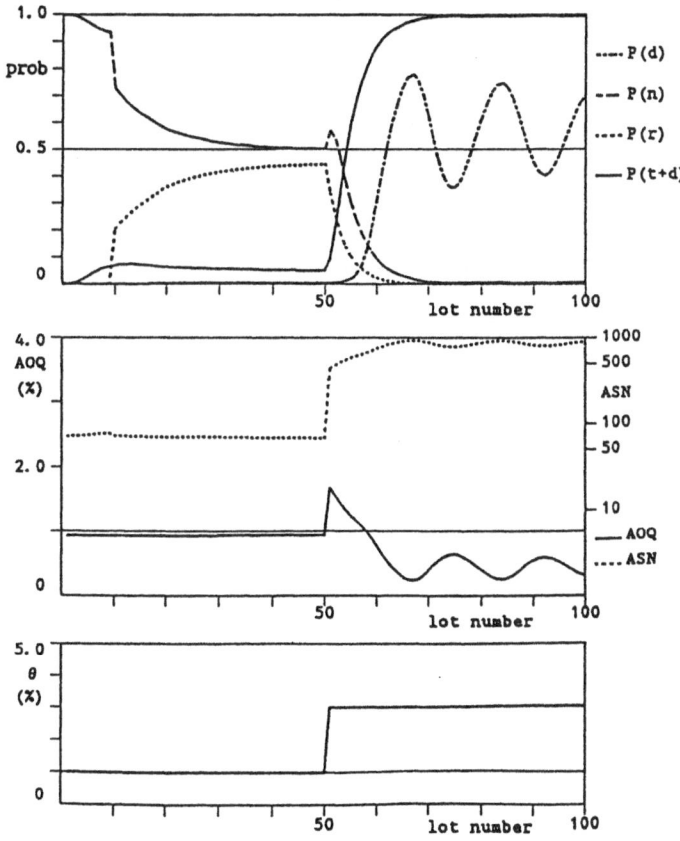

Figure 9: Dynamic features of transitions in case of Fig. 8-(b)

of Figure 9 to 13 are defined to indicate the probability of NI, RI, discontinuation of inspection (DI) and the sum of TI and DI, respectively.

Every curve in Figure 9 is naturally based on the new ISO 3951: 1989(E). But so far as the authors investigated, the present results are quite similar as T.Koyama(p.123) in [4] based on ISO 3951(1979). It seems to us that the revision is almost non-sense, although the procedure "discontinuation" is a little bit modified.

Another illustration is shown in Figure 10, which corresponds to Figure 7-(b) and 7-(a). The thick lines of each kind show the features of the plan specified by Figure 7-(b) and the thin lines show those of Figure 7-(a). From Figures 9 and 10, it will be noticed generally that the outgoing quality level is much smaller than the requested AOQ, when input quality is inferior to AQL. This may mean that the given s.i.p. is too severe.

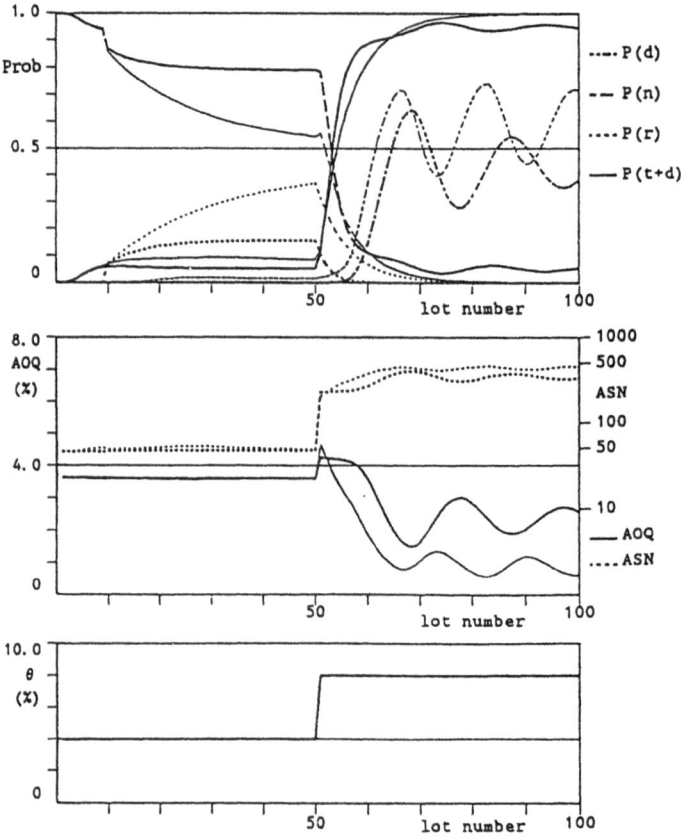

Figure 10: Dynamic features of transitions in case of Fig. 7-(b) and 7-(a)

5.3 PROPERTIES OF ISO 3951 FOR WEIBULL DISTRIBUTIONS

The probabilities of transitions among NI, RI and TI procedures are investigated for Weibull distributions by using a plan shown in Figure 7-(a) of ISO 3951. The dynamic features are shown in Figure 11, where the input quality is assumed to conform the Weibull distribution $W(0.5, 4.0, 3.2)$.

The mean of input quality is set to 4.08 % and AQL = 4.0 %.

The curves are seemed to be almost stable. On the other hand, when the mean of input quality is 4.98 and the other conditions are quite the same as those in Figure 11, the dynamic features are shown in Figure 12. In this case, the transition probability curves are not so stable, comparing with those in Figure 8, because AQL = 4.0% is smaller than the mean 4.98 of input quality. Through these Figures, however, AOQ curves are lower than the value 4.0% of AQL, and moreover the AOQ curve in Figure 12 is lower than the curve in Figure 11.

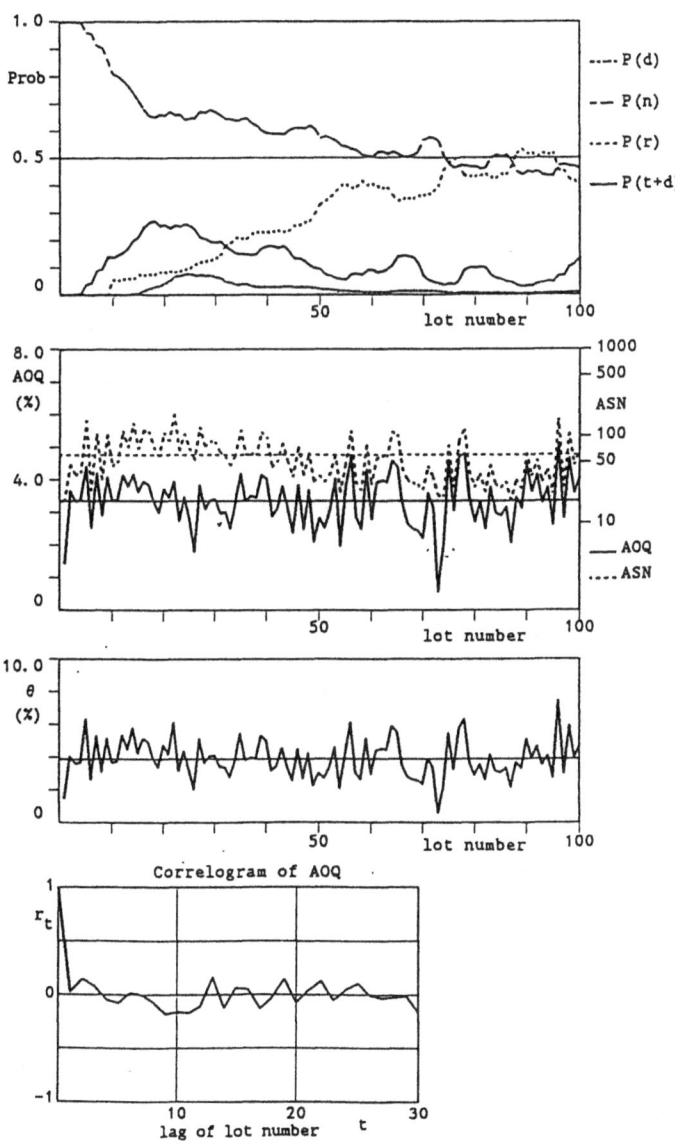

Figure 11: Dynamic features of transitions in case of Fig. 7-(a)
for Weibull distr., quality mean= 4.08%

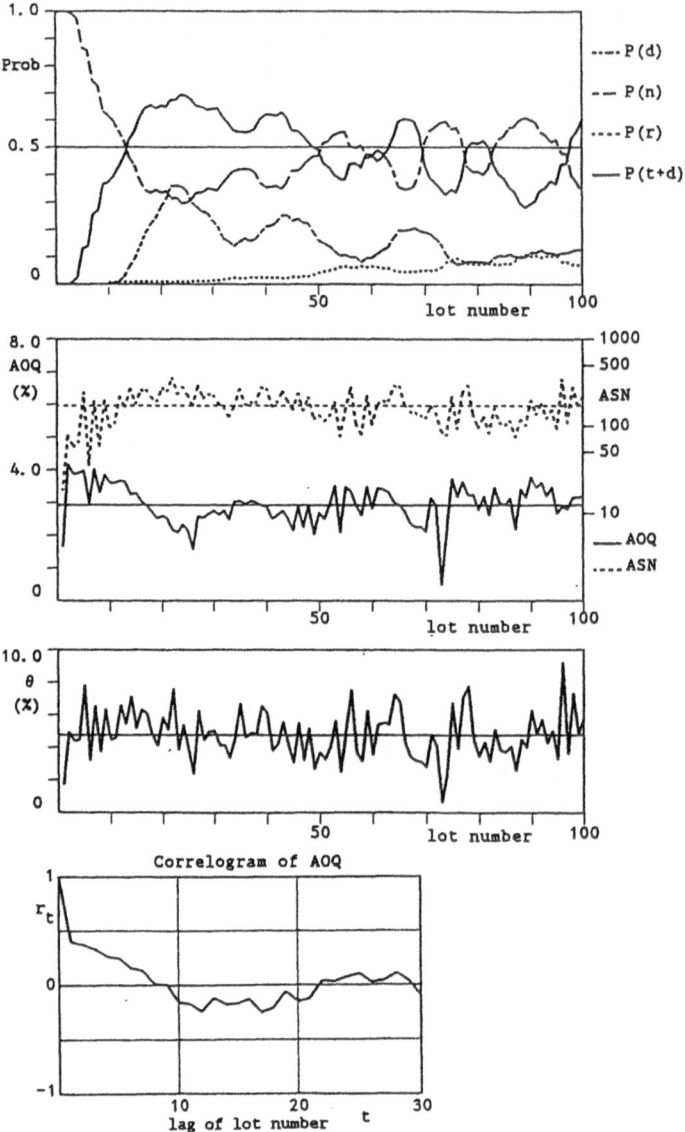

Figure 12: Dynamic features of transitions in case of Fig. 7-(a)
for Weibull distr., quality mean= 4.98%

From these facts, it seems to us again that plans in ISO 3951 have unnecessarily a feature of excessive action, as a whole.

5.4 PROPERTIES OF ISO 3951 FOR UNIFORM DISTRIBUTIONS

Figure 13 shows dynamic curves of the feature of a plan shown in Figure 7-(a) of ISO 3951, as the above section but for the uniform distribution. The mean of input quality is 4.0%. What this result shows is just the same features as shown in the above section.

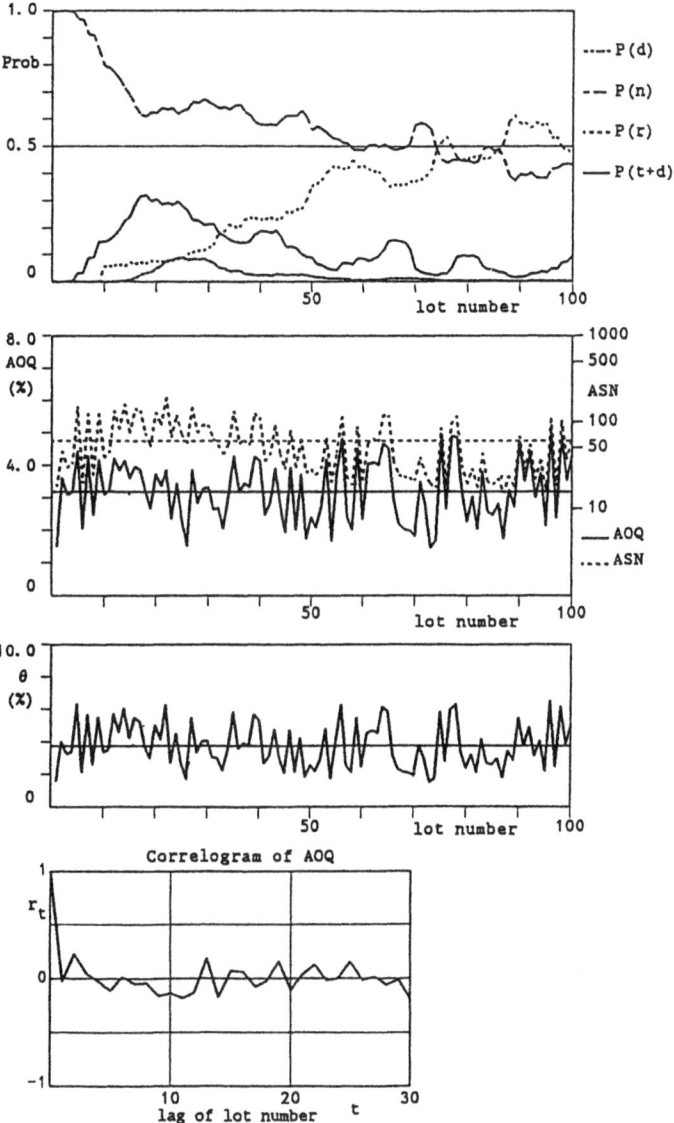

Figure 13: Dynamic features of transitions in case of Fig. 7-(a)
for uniform distr., quality mean= 4.0%

6. CONCLUSION

Micro-computers have been rapidly developing, and many complex sampling inspection plans can be designed and confirmed with their statistical properties by the use of such computers. As a general purpose statistical package, Micro-NISAN system with statistical knowledge functions is quite useful for speedy development of progressive sampling inspection plans. The discussed sampling inspection plans illustrate this.

Regarding single and multi-stage s.i.p.s, more effective plans on life-time test are considered on the reliability of products, e.g. on the basis of the most powerful rank test for Lehmann alternatives, and so on. Specially, regarding the plans involved in ISO 3951, very complicated procedures with the rules are defined. After this, the feature as transitions among NI, RI and TI and the effects of time-lags for the switching rules are to be investigated more in detail by the analysis of auto- and cross- correlations and so on. So far as we studied, however, it seems to us that the plans with statistical inference have clearly something else at the same time, and that in this sense the plans might be sophisticated and essentially to put pressure upon deliverers.

7. REFERENCES

[1] Asano,Ch. et al. (1989): NISAN and Micro-NISAN systems for tools of designing sampling inspection plans, Quality for Progress and Development, pp.590-600, Wiley Eastern.

[2] Kimura,H. and Asano,Ch.(1990): A study on system construction of general purpose statistical package for micro-computers — Micro-NISAN System, Computational Statistics Quarterly, Vol. 6, No. 1.

[3] Asano,Ch. et al.(1984): Developmental studies and evaluation of sampling inspection plans on the basis of a software system, Frontiers in Statistical Quality Control, pp.107-120, Physica-Verlag.

[4] Koyama,T.(1981): Average outgoing quality through sampling systems, Frontiers in Statistical Quality Control, ed. Lenz,H.-J. et al., pp.113-132, Physica-Verlag.

[5] ISO (1989): ISO 3951: Sampling Procedures and Charts for Inspection by Variables for Percent Nonconforming, Second Edition.

On the Design of Exact Single Sampling Plans by Variables

M. Bruhn-Suhr and W. Krumbholz, Hamburg, Germany
H.-J. Lenz, Berlin, Germany

1. Introduction

We deal with sampling by variables in the case of a $N(\mu, \sigma^2)$ distributed quality characteristic X with unknown $\sigma^2 > 0$. Assuming that there are both upper and lower specification limits $U, L\,(L < U)$ the percent defective p is given by

$$p := p_L + p_U =: \psi(\mu, \sigma) \tag{1}$$

with

$$p_L := P(X < L) = \Phi\left(\frac{L - \mu}{\sigma}\right), \tag{2}$$

$$p_U := P(X > U) = \Phi\left(\frac{\mu - U}{\sigma}\right) \tag{3}$$

and Φ denoting the $N(0, 1)$ distribution function. Let

$$\mu_o := \frac{L + U}{2} \tag{4}$$

be the center of the tolerance interval and

$$\sigma_o(p) := \frac{L - U}{2\Phi^{-1}(\frac{p}{2})} \qquad (0 < p < 1). \tag{5}$$

It is evident that

$$
\left.
\begin{aligned}
&\text{(i)} \quad \psi(\mu_o + x, \sigma) = \psi(\mu_o - x, \sigma); \\
&\text{(ii)} \quad \psi(\mu, \sigma) \text{ is increasing in } \mu\,(\mu \geq \mu_o) \text{ for fixed } \sigma > 0; \\
&\text{(iii)} \quad \psi(\mu, \sigma) \text{ is increasing in } \sigma \text{ for fixed } \mu; \\
&\text{(iv)} \quad \text{for each } \sigma\,(0 < \sigma \leq \sigma_0(p)) \\
&\qquad \text{there is exactly one } \mu = \mu(\sigma, p) \geq \mu_o \text{ and exactly one } \mu' = \mu'(\sigma, p) \leq \mu_o \\
&\qquad \text{with } \psi(\mu(\sigma, p), \sigma) = \psi(\mu'(\sigma, p), \sigma) = p; \\
&\text{(v)} \quad \text{for } \sigma_0(p) \text{ defined by (5) we get } \mu'(\sigma_0(p), p) = \mu(\sigma_0(p), p) = \mu_o.
\end{aligned}
\right\} \tag{6}
$$

Figure 1 shows the lines of constant ψ in the μ, σ-plane ("iso-p-lines").
Let $X_1, \dots, X_n$ be a sample on $X\,(n \geq 3)$, and

$$\overline{X} = \frac{1}{n}\sum_i X_i, \qquad S^2 = \frac{1}{n-1}\sum_i (X_i - \overline{X})^2.$$

Let B_r denote the distribution function of the symmetrical beta distribution of the first kind over $(0, 1)$ with parameter r.

Frontiers in Statistical Quality Control 4
Ed. by Lenz et al.
© Physica-Verlag Heidelberg 1992

84

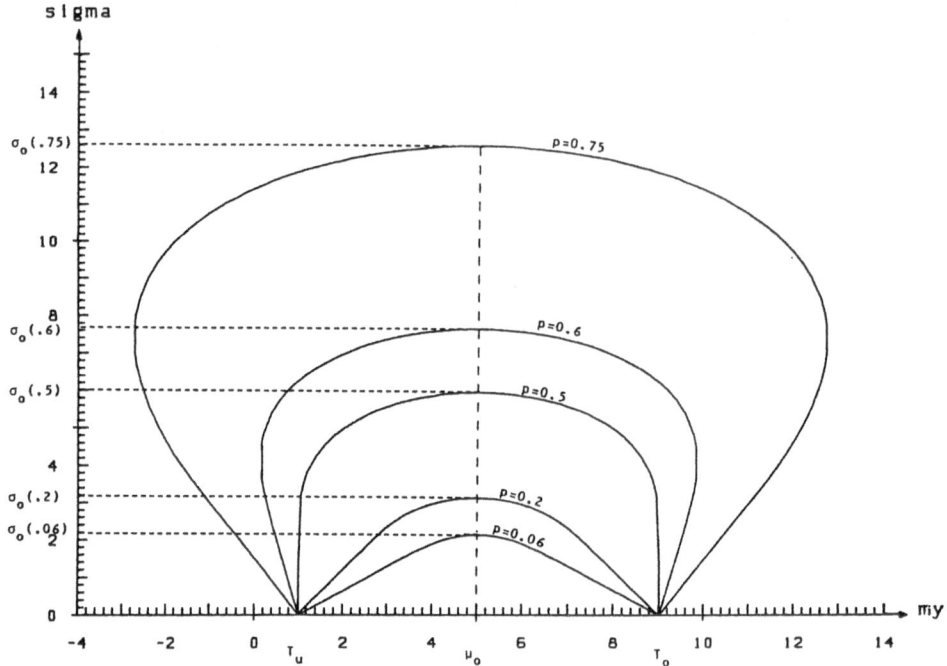

Figure 1: Iso-p-lines

It is well-known (cf. Kolmogorov [4], Lieberman/Resnikoff [5]) that

$$\hat{p}_L := B_{\frac{n}{2}-1}(V), \qquad (7)$$

$$\hat{p}_U := B_{\frac{n}{2}-1}(W), \qquad (8)$$

$$\hat{p} := \hat{p}_L + \hat{p}_U \qquad (9)$$

are the UMVU estimators of p_L, p_U and p respectively, V and W being defined as

$$V := \max\left\{0, \frac{1}{2} - \frac{\sqrt{n}}{n-1}\frac{\overline{X}-L}{2S}\right\},$$

$$W := \max\left\{0, \frac{1}{2} - \frac{\sqrt{n}}{n-1}\frac{U-\overline{X}}{2S}\right\}.$$

Lieberman/Resnikoff [5] proposed the sampling plan (n_1, k_1) based on $\hat{p}$. The lot is accepted if $\hat{p} \le k_1$. Let

$$L_1(\mu, \sigma) := P_{\mu, \sigma}(\hat{p} \le k_1) \qquad (10)$$

denote the OC of (n_1, k_1). By symmetry we get

$$L_1(\mu_0 + x, \sigma) = L_1(\mu_0 - x, \sigma), \qquad (11)$$

and thus for $\mu(\sigma,p),\mu'(\sigma,p)$ defined in (6)(iv) :

$$L_1(\mu'(\sigma,p),\sigma) = L_1(\mu(\sigma,p),\sigma). \tag{12}$$

For fixed p $(0 < p < 1)$ let

$$L_1(\sigma;p) := L_1(\mu(\sigma,p),\sigma) \qquad (0 < \sigma \leq \sigma_0(p)). \tag{13}$$

Remark: Let (n_1', k_1') be the corresponding sampling plan in the case of an one-sided specification limit, which without loss of generality may be assumed to be an upper one U. (n_1', k_1') is based on $\hat{p}_U$ defined in (8). The lot is accepted if $\hat{p}_U \leq k_1'$. The OC of $(n_1', k_1') = (n, k)$ satisfies

$$
\begin{aligned}
L_1'(\mu,\sigma) &:= P_{\mu,\sigma}(\hat{p}_U \leq k) \tag{14} \\
&= F_{n-1,\delta}((n - 1)(2B_{\frac{n}{2}-1}^{-1}(k) - 1)) \\
\text{with } \delta &:= \sqrt{n}\frac{\mu - U}{\sigma} \tag{15}
\end{aligned}
$$

and $F_{r,\delta}$ denoting the distribution function of the noncentral t distribution with r degrees of freedom and noncentrality parameter δ.

If $\sigma \ll \sigma_0(p)$, we have $\hat{p} \approx \hat{p}_U$ and $\hat{p} \approx \hat{p}_L$ a.s., respectively. Thus, by the symmetry (11), (12) together with (3), (10), (14), (15) we get

$$
\begin{aligned}
L_1(\sigma;p) &\approx L_1'(\mu(\sigma,p),\sigma) \tag{16} \\
&= F_{n-1,\sqrt{n}\Phi^{-1}(p)}((n - 1)(2B_{\frac{n}{2}-1}^{-1}(k) - 1)) \quad (\sigma \ll \sigma_0(p)).
\end{aligned}
$$

Numerical investigations by Resnikoff [6] and recent calculations by Baillie [1] show that for p being fixed $L_1(\sigma;p)$ varies only over a narrow band as σ varies over the whole range $0 < \sigma \leq \sigma_0(p)$. Therefore it seems to be reasonable to use the approximation (16) of $L_1(\sigma;p)$ for all $\sigma \in (0,\sigma_0(p)]$. The corresponding plan $(\bar{n}_1, \bar{k}_1)$, which is obtained by (14) - (16) in the same way as the one-sided plan (n_1', k_1'), is called the one-sided approximation of (n_1, k_1). Lieberman/Resnikoff [5] consequently use $(\bar{n}_1, \bar{k}_1)$ instead of (n_1, k_1) itself. A formula for the exact OC of (n_1, k_1) was proved in [3].

Bruhn-Suhr/Krumbholz [2] proposed the sampling plan (n_2, k_2) based on the estimator

$$p^* := \psi(\overline{X}, S) \tag{17}$$

of p, which is essentially the ML estimator. The lot is accepted if $p^* \leq k_2$.

Remark: The plan (n_2, k_2) can be approximated by the corresponding one-sided approximation plan $(\bar{n}_2, \bar{k}_2)$ like (n_1, k_1) by $(\bar{n}_1, \bar{k}_1)$. $(\bar{n}_2, \bar{k}_2)$ is obtained in the same way as the corresponding one-sided plan (n_2', k_2') based on $p_U^* := \Phi(\frac{\overline{X}-U}{S})$. The one-sided plans (n_1', k_1') and (n_2', k_2') are equivalent, if $n_1' = n_2'$ and

$$k_2' = \Phi\left(\frac{n - 1}{\sqrt{n}}\left(2B_{\frac{n}{2}-1}^{-1}(k_1') - 1\right)\right).$$

A formula for the exact OC $L_2(\mu,\sigma) := P_{\mu,\sigma}(p^* \le k_2)$ of (n_2,k_2) was proved in [2]. This formula is considerably less complicated than the above mentioned formula for the exact OC of the Lieberman-Resnikoff-plan (n_1,k_1).

The formulas for the OC of (n_i,k_i) $(i=1,2)$ allow several applications in process control as well as in acceptance sampling. As an example we deal with the "classical" problem of determining the two-sided sampling plan $(n,k) = (n_i,k_i)$ $(i=1$ or $2)$ in the case of given AQL p_1, LQ p_2, and type I and II errors α and β $(0 < p_1 < p_2 < 1; 0 < \beta < 1 - \alpha < 1)$. The OC L of (n,k) has to satisfy the conditions

$$
\left.
\begin{array}{ll}
\text{(i)} & \displaystyle\min_{\substack{\mu,\sigma \\ \psi(\mu,\sigma)\le p_1}} L(\mu,\sigma) \ge 1-\alpha \\[4ex]
\text{(ii)} & \displaystyle\max_{\substack{\mu,\sigma \\ \psi(\mu,\sigma)\ge p_2}} L(\mu,\sigma) \le \beta
\end{array}
\right\} \tag{18}
$$

with minimal sample size n. Using certain monotony properties of the OC of (n_i,k_i) $(i=1,2)$ we see that (18)(i), (ii) are equivalent to

$$
\left.
\begin{array}{ll}
\text{(i)} & \displaystyle\min_{0<\sigma\le\sigma_0(p_1)} L(\sigma;p_1) \ge 1-\alpha \\[4ex]
\text{(ii)} & \displaystyle\max_{0<\sigma\le\sigma_0(p_2)} L(\sigma;p_2) \le \beta.
\end{array}
\right\} \tag{19}
$$

In [2] and [3] an approach called "successive improvement" was developed. Starting with the one-sided approximation plan $(\bar{n}_i,\bar{k}_i)$ determined using (16) it allows the evaluation of the plan (n_i,k_i) which satisfies (19) (i), (ii) exactly.

Example 1

$$L = 1.0,\ U = 9.0,\ p_1 = 0.01,\ p_2 = 0.03,\ \alpha = \beta = 0.10$$

i	$(\bar{n}_i,\bar{k}_i)$	$\min\limits_{\sigma} L_i(\sigma;p_1)$	$\max\limits_{\sigma} L_i(\sigma;p_2)$	(n_i,k_i)
1	$(108; 0.0168)$	0.8968	0.1060	$(113; 0.0168)$
2	$(108; 0.0175)$	0.8845	0.0995	$(115; 0.0178)$

Remark: Numerical investigations showed that there is no need for $L_i(\bullet;p)$ $(i=1,2)$ to have either its maximum or its minimum in the region $\sigma \ll \sigma_0$ or in $\sigma = \sigma_0(p)$. Furthermore, $L_i(\bullet;p)$ $(i=1,2)$ may have several local maxima or minima. Therefore the computation of (n_i,k_i) $(i=1,2)$ needs a reasonable amount of CPU time. Because the UMVU estimator $\hat{p}$ is superior to the (quasi) ML estimator p^*, the exact Lieberman-Resnikoff-plan (n_1,k_1) is more efficient (in the sense of Fisher) than (n_2,k_2). Therefore in example 1 the sample size of (n_1,k_1) is a little lower than that of (n_2,k_2). On the other hand, numerous calculations which were done using our former programs showed that the little loss of sample size of (n_2,k_2) is overcompensated by a saving of about 80 % of computing time. This is caused by the considerably simpler formula for the OC of (n_2,k_2). Therefore, in the algorithm which will be presented in the next section, we only deal with exact plans (n_2,k_2).

2. The Algorithm

Given AQL p_1, LQ p_2, and type I and II errors α and β $(0 < p_1 < p_2 < 1; 0 < \beta < 1 - \alpha < 1)$, we want to determine the plan $(n,k) = (n_2, k_2)$ with the OC $L(\mu,\sigma) = L_{(n,k)}(\mu,\sigma) = L_2(\mu,\sigma)$ satisfying (19) (i),(ii). Setting

$$M_1(n,k) := \min_{0<\sigma\leq\sigma_0(p_1)} L_{(n,k)}(\sigma; p_1), \tag{20}$$

$$M_2(n,k) := \max_{0<\sigma\leq\sigma_0(p_2)} L_{(n,k)}(\sigma; p_2), \tag{21}$$

(19) (i),(ii) is equivalent to

$$M_1(n,k) \geq 1 - \alpha, \tag{22}$$

$$M_2(n,k) \leq \beta. \tag{23}$$

In order to obtain an initial value of the sample size n we determine the sample size n^* of the usual approximation plan (n^*, k^*) of the corresponding one-sided plan (n_2', k_2') (cf. [7], p.184-185). Using the regression line

$$n \approx 1.0315\,n^* + 3.036$$

which was obtained by evaluation of 23 plans, we define the initial value of n

$$n_0 := [1.0315\,n^* + 3.036], \tag{24}$$

with $[x]$ denoting the greatest integer $\leq x\,(x \in \mathbb{R})$. Using $n := n_0$, we determine the solution $k(n)$ of the equation $M_1(n,k) = 1 - \alpha$ in the next step and afterwards $M_2(n, k(n))$. If $M_2(n, k(n)) \leq \beta$ then the algorithm backtracks ($"n := n - 1"$) else a branch to $n := n + 1$ is performed.

The following algorithm gives a more detailed description of the procedure using a pseudocode notation.

ALGORITHM VARIAB

input : $p_1, p_2, \alpha, \beta, L, U, itmax$
output : n, k
begin

```
        i := 0                                          {initializing}
        istop := .false. ; iback := .false.
        (n,k) := proxi(p_1,p_2,\alpha,\beta)            {find proxi of (n,k)}
        M_1 := M_1(n,k); M_2 := M_2(n,k)                {compute OC-values}
        if [(M_1 \geq 1 - \alpha) and (M_2 \leq \beta)] then iback := .true.    {feasibility check}
        endif
```

```
repeat
      i := i + 1
      solve M₁(n,k) = 1 − α for k ∈ (0,1)                    {find new k}
      M₂ := M₂(n,k)                                           {compute OC-values}
      if (M₂ ≤ β) then                                        {feasibility check}
         if (iback) then n := n − 1                           {backtrack}
                  else istop := .true.                        {(n,k) found}
      endif
      elseif (M₂ > β) then n := n + 1                         {branch to n + 1 }
                    iback : = .false.
      endif
until [(istop) or (i > itmax)]
end
```

The algorithm is implemented in FORTRAN 77 using the F 77 LE - 32 compiler of Lahey and DOS 3.3 on a PS-2/80 IBM-PC. A typical run of the algorithm is given in the following example.

Example 2

$$L = 1.0, U = 9.0, p_1 = 0.01, p_2 = 0.06, \alpha = 0.01, \beta = 0.10, n^* = 58$$

i	n_i	$k_i = k(n_i)$	$M_1(n_i, k_i)$	$M_2(n_i, k_i)$
0	62	0.0347	0.9900	0.1099
1	63	0.0344	0.9900	0.1050
2	64	0.0341	0.9900	0.1002
3	65	0.0339	0.9900	0.0957

The trajectory of a run.$(n,k) = (65, 0.0339)$, $C(65) = 194$.

The time complexity $C(n)$ of the algorithm is mainly influenced by the time for solving the equations $M_1(n,k) = 1 - \alpha$ and the determination of $M_2(n, k(n))$. This is caused by a complex integral in the formula for the exact OC of (n_2, k_2) (cf. [2]). The upper limit of these integrals is proportional to n. This explains the nearly linear behaviour of the CPU-time over n for all sample sizes $20 \leq n \leq 80$ in figure 2.

Finally, we give several examples of plans $(n_2, k_2) = (n, k)$. These are listed below. Furthermore, in Table 1 we give the sample size $\bar{n}_2 = \bar{n}$ of the corresponding one-sided approximation $(\bar{n}_2, \bar{k}_2)$ and the "rate of improvement" $\frac{n - \bar{n}}{\bar{n}} 100$ in percent.

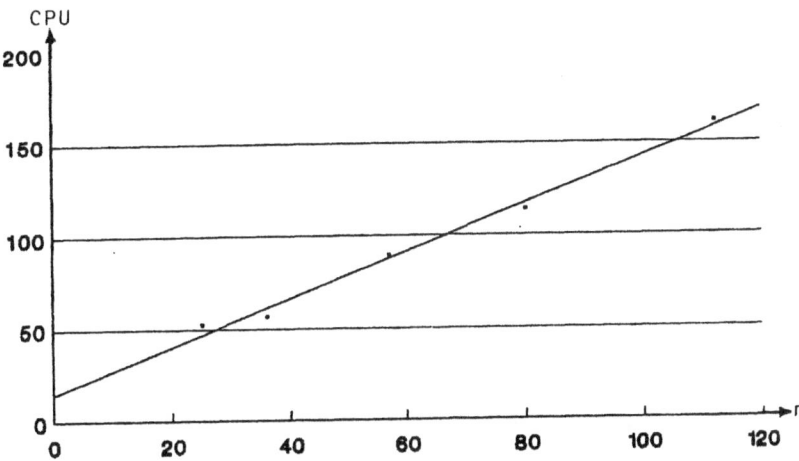

Figure 2: CPU time in s, $\alpha = \beta = 0.01, 0.025, 0.05, 0.1, 0.15, p_1 = 0.01, p_2 = 0.06, L = 1, U = 9$

L	U	p_1	p_2	α	β	(n, k)	$\bar{n}$	$\frac{n-\bar{n}}{\bar{n}}100$
1	9	0.01	0.06	0.10	0.10	$(36, 0.0264)$	33	9.09
1	9	0.01	0.06	0.01	0.01	$(112, 0.0265)$	105	6.67
1	9	0.01	0.06	0.01	0.10	$(65, 0.0339)$	58	12.07
1	9	0.01	0.06	0.10	0.01	$(71, 0.0205)$	68	4.41
1	9	0.01	0.06	0.025	0.10	$(54, 0.0316)$	48	12.50
1	9	0.01	0.06	0.025	0.025	$(80, 0.0266)$	75	6.67
1	9	0.01	0.06	0.10	0.20	$(25, 0.0311)$	21	19.05
1	18	0.01	0.06	0.10	0.10	$(36, 0.0264)$	33	9.09
1	18	0.01	0.06	0.01	0.01	$(112, 0.0265)$	105	6.67
1	27	0.01	0.06	0.10	0.10	$(36, 0.0264)$	33	9.09
1	27	0.01	0.06	0.01	0.01	$(112, 0.0265)$	105	6.67
1	9	0.005	0.03	0.10	0.10	$(53, 0.0130)$	49	8.16
1	9	0.005	0.03	0.01	0.10	$(96, 0.0169)$	88	9.09
1	9	0.005	0.03	0.10	0.05	$(69, 0.0117)$	64	7.81
1	9	0.005	0.03	0.025	0.10	$(79, 0.0157)$	72	9.72
1	9	0.005	0.02	0.10	0.10	$(96, 0.0104)$	90	6.67
1	9	0.01	0.03	0.10	0.10	$(115, 0.0178)$	108	6.48
1	9	0.01	0.04	0.10	0.10	$(67, 0.0209)$	62	8.06
1	9	0.01	0.05	0.10	0.10	$(47, 0.0238)$	43	9.30

Table 1. the sample size $\bar{n}_2 = \bar{n}$ of the corresponding one-sided approximation $(\bar{n}_2, \bar{k}_2)$ and the "rate of improvement" $\frac{n-\bar{n}}{\bar{n}}100$

3. References

[1] D.H. Baillie (1987) :   *Multivariate Acceptance Sampling.*
Frontiers in Statistical Quality Control 3, 83-115. Ed. by Lenz et al., Heidelberg : Physika

[2] M. Bruhn-Suhr, W. Krumbholz (1990) :   *A new variables sampling plan for normally distributed lots with unknown standard deviation and double specification limits.*
Statistical Papers, Vol. 31, Number 3 , 195-207.

[3] M. Bruhn-Suhr, W. Krumbholz (1989) :   *Exact Two Sided Lieberman-Resnikoff Sampling Plans.*
Discussion Papers in Statistics and Quantitative Economics Nr. 36, Universität der Bundeswehr Hamburg (A revised version will appear in the Statistical Papers.).

[4] A.N. Kolmogorov (1953) :   *Unbiased Estimates.*
American Mathematical Society Translations 98, 28pp.

[5] G.J. Lieberman and G.J. Resnikoff (1955) :   *Sampling Plans for Inspection by Variables.*
Journal of the American Statistical Association 50, 457-516.

[6] G.J. Resnikoff (1952) :   *A New Two Sided Acceptance Region for Sampling by Variables.*
Technical Report Nr. 8, Applied Mathematics and Statistics Laboratory, Stanford University

[7] W. Uhlmann (1982) :   *Statistische Qualitätskontrolle.*
2nd edn. Stuttgart: Teubner

The Pitfall of Acceptance Sampling by Variables

E. v. Collani, Würzburg, Germany

1. INTRODUCTION

To introduce the problem consider the following citations taken
from well-known sources of statistical quality control:
- "The term "acceptance sampling" relates to the acceptance or rejection of a product or process on the basis of sampling inspection.
  ... Sampling plans, which specify sample size and acceptance criteria, are fundamental to acceptance sampling. Such plans may be based
  on a simple dichotomous classification of conformance or nonconformance of a quality characteristic to specified criteria (attributes
  plans) or on a comparison of statistics computed from quantitative
  measurements to numerical criteria developed from the specifications
  and from assumptions about the shape and nature of the distribution
  of individual measurements (variables plans)." (Kotz, Johnson (eds):
  Encyclopedia of Statistical Sciences, Vol. 1, pp. 12, 13.)

Accordingly there are two different purposes related to acceptance
sampling, namely judging incoming lots with respect to their quality
and judging a process with respect to its process parameters respectively. This paper is mainly concerned with the former one, i.e.
acceptance sampling by variables to control the fraction nonconforming
in a lot. Montgomery describes them in the following way:
- "Variables sampling plans specify the number of items to be sampled
  and the criterion for sentencing lots when measurements data are
  collected on the quality characteristic of interest. These plans are
  generally based on the sample average and sample standard deviation
  of the quality characteristic. When the distribution of the quality
  characteristic in the lot or process is known, variables sampling
  plans that have specified risks of accepting and rejecting lots of
  given quality may be designed." (Montgomery: Introduction to Statistical Quality Control, p. 431.)

Clearly variables sampling plans are used to test the fraction nonconforming in given lots, and frequently they are designed to have specified consumer's and producer's risks for given quality levels.
In what follows it is shown that the above stated purpose is not at

all compatible with the usual practice of variables sampling. To demonstrate this the simplest case of a univariate quality characteristic distributed normally with known standard deviation is chosen. Of course, the same holds in the general case, too. Schilling describes, how to proceed in this case:

- "$\bar{X}$ Method. The simplest application of variables plans for proportion nonconforming is when a single specification limit is involved and the standard deviation is known. In this case, a straightforward procedure, which we shall call the $\bar{X}$ method, may be employed. It requires that the sample size and an acceptance constant k be specified and that σ be known. An acceptance limit A for $\bar{X}$ is set at distance kσ within the specification limits. The procedure, then, is as shown in Table 10-1.

TABLE 10-1 $\bar{X}$ Method

Lower specification limit	Upper specification limit
1. Set A = L + kσ.	1. Set A = U − kσ.
2. Select a random sample of size n.	2. Select a random sample of size n.
3. Compute $\bar{X}$.	3. Compute $\bar{X}$.
4. If $\bar{X} \geq A$, accept the lot;	4. If $\bar{X} \leq A$, accept the lot;
if $\bar{X} < A$, reject the lot.	if $\bar{X} > A$, reject the lot.

."

(Schilling: Acceptance Sampling in Quality Control. p. 229.)

The reason for applying a variables plan instead of an attributes plan is given for instance by Duncan:

- "When a quality characteristic is measurable on a continuous scale and is known to have a distribution of a specific type, say it is known to be normally distributed, it may be possible to use as a substitute for an attributes sampling plan a sampling plan based on sample measurements ... These variables sampling plans have the primary advantage that the same operating characteristic curve can be obtained with a smaller sample than is required by an attributes plan."

Some lines further Duncan notes a difficulty with respect to variables plans:

- "All variables plans, however, assume some way of going from a mean and standard deviation to a fraction defective. If the items are not normally distributed, estimates of the fraction defective based on the mean and standard deviation will not be the same as if the items were normally distributed."

Of course, the normality assumption does never hold for any given lot.
Duncan overcomes this difficulty by the following assumption:
- "In practice, if a lot is large, we can talk about being normally
  distributed without serious difficulty. This may not be true for
  small lots. We shall assume here that if a small lot has a bell-
  shaped distribution, with the same γ coefficients as the normal distributions, viz. $\gamma_1'=0$, $\gamma_2'=0$, it will be satisfactory to consider it
  as 'normally distributed'." (Duncan: Quality Control and Industrial
  Statistics, pp. 219-221.)

In the following sections it is shown that the last assumption is
false i.e. the discrete and finite distribution of the quality characteristic within a given lot cannot be approximated sufficiently well
by a normal distribution in order to "go from the mean and standard
deviation to the fraction defective" being the quantity of primary interest.

2. ACCEPTANCE SAMPLING BY VARIABLES

- "Acceptance sampling is a major component of quality control. In acceptance sampling, inspection is performed on a shipment or lot of
  incoming materials. Samples are taken from this shipment, and certain quality characteristics of the units are inspected. After this
  inspection, a decision is made concerning the shipment. Usually the
  decision is either to accept or reject the lot.
  One purpose of acceptance sampling is to determine a course of action
  (i.e., to accept or reject). Another purpose is to prescribe a procedure that, if applied to a series of lots, will give a specified
  risk of accepting a lot of given quality level. It should be emphasized that the purpose of acceptance sampling is not to estimate or
  to control the quality of the product. Controlling quality is the
  purpose of control charts." (Banks: Principles of Quality Control,
  p. 293.)

As stated in the above quotation the risks of a false decision for
given quality level play a decisive role in designing a sampling plan.
These risks can be taken from the socalled operating characteristic
(OC) curve that "shows the probability of accepting a lot submitted
for inspection for a range of fraction nonconforming values" (Banks,
p. 295).
According to Table 10-1 (Schilling) a variable sampling plan for the
situation described in the previous section is given by a pair of numbers (n,k) with:
n = sample size,

k = acceptance number with k>0,

and the following rule:

Select a random sample of size n given by $(x_1', \ldots, x_n')$ out of the given lot $(x_1, x_2, \ldots, x_N)$ of size N, where the i-th item in the lot is

$$\left.\begin{array}{l} \text{conforming iff } x_i \le U, \text{ and} \\ \text{nonconforming iff } x_i > U. \end{array}\right\} \qquad (1)$$

Compute

$$\bar{x} = \frac{1}{n} \sum_{i=1}^{N} x_i' \qquad (2)$$

with $x_1', \ldots, x_n'$ denoting the sample observations.

Then

$$\left.\begin{array}{l} \text{if } \bar{x} \le U - k\sigma \quad \text{accept the lot, and} \\ \text{if } \bar{x} > U - k\sigma \quad \text{reject the lot.} \end{array}\right\} \qquad (3)$$

Obviously the above defined sampling plan determines a course of action and therefore meets the first aim listed by Banks. Hence it remains to check whether the second aim mentioned by Banks is met too, i.e. whether a variables sampling plan (n,k) assures that "good" lots are rejected, and "bad" lots are accepted only with small, prescribed probabilities (producer's risk and consumer's risk respectively). To this end we have to look at the OC curve of a variables sampling plan, i.e. the probability of accepting a given (i.e. shipped) lot with fraction nonconforming p.

3. THE OPERATING CHARACTERISTIC FUNCTION

The traditional way of computing the OC curve which can be found in almost any textbook on quality control is based on the assumption stated by Duncan that in general the normality assumption holds sufficiently well. We briefly sum up the proceeding:

Let X denote the univariate quality characteristic of interest, and X_i (i=1,2,...,N) the quality characteristic of the i-th item in the lot submitted for inspection.

It is assumed that $\{X_i\}_{i=1,2,\ldots,N}$ is a set of independent and normally distributed random variables with known standard deviation σ(=const.) and unknown expectation μ:

$$X_i \sim N(\mu, \sigma^2) \quad \text{for } i=1,2,\ldots,N. \qquad (4)$$

Hence:

Prob(the i-th item in the lot is nonconforming) =

$$\text{Prob}(X_i > U) = 1 - \text{Prob}(X_i \le U) = 1 - \Phi(\frac{U-\mu}{\sigma}) =: \pi, \qquad (5)$$

with Φ denoting the distribution function of a standardized normal random variable.

Obviously there is a one-to-one relation between μ and π, moreover we have

$$\mu = U - \sigma \Phi^{-1}(1-\pi). \tag{6}$$

π is the probability of drawing a nonconforming item out of the set $\{X_i\}_{i=1,2,..}$ used to represent the lot in question. We call π the defect probability.

Applying a variables sampling plan (n,k) to this situation means to perform a test of hypothesis on the process parameter μ or equivalently on the defect probability π:

$$H_0 : \mu \leq U \quad \text{or equivalently} \quad H_0' : \pi \leq \frac{1}{2}. \tag{7}$$

Not rejecting the nullhypothesis is tantamount to accept the given lot, with

$$
\begin{aligned}
\Pr(\text{not rejecting } H_0' | \pi) &= \Pr(\bar{X} \leq U - k\sigma/\sqrt{n} | \pi) = \\
&= \Pr(\frac{\bar{X}-\mu}{\sigma} \sqrt{n} \leq -k + \frac{U-\mu}{\sigma} \sqrt{n} | \pi) = \\
&= \Phi(-k + \frac{U-\mu}{\sigma} \sqrt{n}) = \\
&= \Phi(-k + \sigma[\Phi^{-1}(1-\pi)]\sqrt{n}). \tag{8}
\end{aligned}
$$

Let p denote the actual fraction nonconforming in the lot submitted for inspection. Then the fraction nonconforming p is identified with the defect probability π in the above given model, and based on the assumption made at the beginning, it is concluded that the following relation holds:

$$\Pr(\text{accepting the lot} | p) \approx \Phi(-k + \sigma[\Phi^{-1}(1-p)]\sqrt{n}). \tag{9}$$

The question to be investigated is whether or not (9) holds sufficiently well, i.e. whether the right hand side of (9) can be used as an approximation of the left hand side at least for large lots or bell shaped distributions of the quality characteristic.

We investigate this question by means of the most simple case, which at the same time is a rather important one, namely lots with fraction nonconforming equal to zero. Clearly a lot containing no nonconforming item at all (p=0) constitutes a good lot, and one would expect that the OC curve for any reasonable sampling plan adopts the value 1 in this case, where the OC function L of an acceptance sampling plan (n,k) is defined by:

L(p) = Prob(accepting a <u>given lot</u> with fraction nonconforming p,
when using the acceptance sampling plan (n,k)), (10)

i.e. L(p) is the left hand side of (9).

Obviously, the right hand side fulfills the above expressed expectation, i.e. for p=0 the right hand side takes the value 1.

Next we have to look at the left hand side of (9) given by L(0). Using the model $\{X_i\}_{i=1,2,..,N}$ introduced at the beginning for the lot in question, we note that there are, of course, infinitely many different events leading to a lot containing no item nonconforming at all. Any subset of the event

$$\{X_i \leq U \quad \text{for } i=1,2,..,N\} \tag{11}$$

leads to a faultless lot, i.e. p=0.

Consider the following two different subsets:

Case 1:

$$A: = \{U - k\sigma < X_i \leq U \quad \text{for } i=1,2,..,N\} \tag{12}$$

with

$$\text{Prob}(A) = [\Phi(\frac{U-\mu}{\sigma}) - \Phi(\frac{U-\mu}{\sigma} - k)]^N > 0. \tag{13}$$

Obviously we have p=0 in the case that A has occurred, and we have independently of the sample size n, the lot size N, and the shape of the distribution of the quality characteristic within the lot respectively:

$$L_A(0) = \text{Prob}(\bar{x} \leq U - k\sigma | U - k\sigma < x_i \leq U, \quad i=1,2,..;N) = 0, \tag{14}$$

where the index A denotes the occurrence of the event A given by (12). Hence in the case that A has occurred, faultless lots are rejected with probability 1 in the case that a variables sampling plan is used.

Remark: The above described situation may well arise and lead to severe differences. Consider for instances the case that the producer removes all items with $x_i > U$ prior to the shipment by means of a screening procedure, but the consumer applies nevertheless a variables sampling plan (n,k).

Case 2:

$$B: = \{X_i \leq U - k\sigma \quad \text{for } i=1,2,..,N\} \tag{15}$$

with

$$\text{Prob}(B) = [\Phi(\frac{U-\mu}{\sigma} - k)]^N > 0. \tag{16}$$

Again we have p=0 in the case that B occurred, but now we get:

$$L_B(0) = \text{Prob}(\bar{x} \leq U - k\sigma | x_i \leq U - k\sigma, \quad i=1,2,..,N) = 1, \tag{17}$$

i.e. in this case the lot is accepted with certainty independently of the sample size n, the lot size N and the shape of the distribution

respectively. With (14) and (17) we have obtained the following (trivial) result for the OC curve:

$$0 \leq L(0) \leq 1. \tag{18}$$

Nontrivial is the fact that (18) is sharp in the sense that it cannot
be improved as both the lower and the upper bound for L(0) can be
adopted.

Moreover we can conclude that L(0) can take almost any value in [0,1]
solely depending on the particular realization of the random vector
$(X_1,X_2,..,X_N)$, i.e. of the measurements $(x_1,x_2,..,x_N)$ in the lot submitted for inspection. And, of course, this result holds analogously
for any value of the lot fraction nonconforming $p \neq 0$ too.

Conclusions:

1. Using the approximation (9) in order to design a variables plan for
   lot-by-lot acceptance sampling makes no sense at all.

2. The above results are independent of the lot size N and the shape
   of the distribution of $\{x_1,x_2,..,x_N\}$.

3. Even more striking: The result is also independent of the sample
   size n, i.e. even in the case of 100% inspection there is no improvement in principle.

Hence, using a variables plan for lot-by-lot inspection of incoming
lots means to accept the fact that there is no protection against
false decisions as the corresponding risks are uncontrollable. By the
way there are many authors on quality control who evidently didn't
feel happy about the usual practice of variables plans in lot-by-lot
inspection. For instance Schilling writes in the Juran (1974) Quality
Control Handbook:

- "Use of these plans in incoming inspection should be restricted to
  product from known and trusted suppliers with a confirmed history of
  a reasonably stable process steadily producing products with a known
  shape of distribution." (Quality Control Handbook, 3rd ed., J.M.Juran,
  F.M.Gryna, R.S.Bingham (eds.): E.G.Schilling: Sampling by Variables.
  Section 25, p. 25-4.)

4. TYPE B OPERATING CHARACTERISTIC FUNCTION

As mentioned in the introduction variables plans are used for two
different purposes. In the preceding sections it is shown that variables plans should not be used for lot-by-lot acceptance sampling. If
variables plans are used for an indirect process control, the sampling

plan has to be looked upon as a means of testing a hypothesis concerning some process parameter. The complement of the power curve for the corresponding statistical test is - following the terminology of Dodge - called Type B Operating Characteristic curve.

Let $L(\pi)$ denote the Type B OC function for the simple example considered in this paper with the defect probability π being the process parameter of interest. Clearly the defect probability π is equal to the expected fraction nonconforming in a lot given by the random vector $(X_1, X_2, \ldots, X_N)$.

Let H_0' be the hypothesis to be tested with respect to π , then it follows from (6) that there exists a hypothesis with respect to the process expectation μ being equivalently to H_0'.

In the relevant literature the Type B OC function for the example considered here is given by

$$L(\pi) = Pr(\bar{X} \leq U - k\sigma/\sqrt{n} \mid \pi) = \Phi(-k + \sigma[\Phi^{-1}(1-\pi)]\sqrt{n}) \qquad (19)$$

corresponding to the hypothesis H_0' or equivalently H_0 given in (7). Clearly both of them are everything else than relevant for the aim of controlling the process, i.e. controlling a target value of the process mean which should be in general far below the upper specification limit U. Hence a Type B OC curve based on (7) seems to be rather incompatibel with the above stated purpose. Of course, the situation is different if there is a second purpose pursued when applying a variable sampling plan, namely not only to decide on the quality of the underlying process - described by the defect probability π - but also to decide on the quality of the shipped lot - given by p. But in this case one has to take into account the fact that the sample $(x_1', \ldots, x_n')$ is taken not from a normal population but from the finite set of realizations $(x_1, x_2, \ldots, x_N)$ representing the lot.

5. CONCLUSIONS

Variables sampling plans, as proposed and recommended in many textbooks on quality control do not (and even worse, cannot) keep what is promised by the authors and what is believed by practitioners. The simple reason for the striking difference between claim and reality is the obvious fact that an expectation of a random variable (here $\pi = E[P]$) is an entirely different quantity than one particular realization of the random variable (here p the lot fraction nonconforming). The one-to-one relation between the expectation μ of the normal quality characteristic X and the defect probability π does not at all establish a similar relation between μ and p being a realization of P. This simple observation has far-reaching consequences for statistical quality control as well as for industrial quality control. In statistical quality control there exist a great number of papers and books investigating variables sampling plans based on a false proposition. In industrial quality control variables sampling plans are widespread because of the false belief that they give equal protection as attributes sampling plans but require smaller sample size.

Let me conclude with a descriptive example due to Mrs. S. Betz, secretary of our department, illustrating the inaptness of lot acceptance sampling by variables: Imagine that you have ordered a batch of cans of green colour but received only cans with blue and yellow. In order to decide upon acceptance or rejection a sample out of a number of randomly chosen cans is taken and mixed, where the mixing procedure represents the computation of the mean $\bar{x}$. Mixing yellow and blue results in green and hence the batch is accepted.

6. REFERENCES

[1]   BANKS, J. (1989): Principles of Quality Control. New York: Wiley & Sons.

[2]   BOWKER,A.H. and GOODE, H.P. (1952): Sampling Inspection by Variables. New York: McGraw-Hill.

[3]   COLLANI, E.v. (1991): A Note on Acceptance Sampling for Variables, Metrika, Vol. 38, in press.

[4]   DUNCAN, A.J. (1965): Quality Control and Industrial Statistics. Homewood: Richard D.Irwin (3rd ed.).

[5]   JURAN, J.M., GRYNA, F.M. and BINGHAM, R.S., eds., (1974): Quality Control Handbook, 3rd. ed., New York: McGraw-Hill.

[6]   KOTZ, S. and JOHNSON, N.L., eds., (1982): Encyclopedia of Statistical Sciences, Vol. 1, pp. 12/13. New York: Wiley & Sons.

[7]   LIEBERMANN, G.J. and RESNIKOFF, G.J. (1955): Sampling Plans for Inspection by Variables. Journal of the American Statistical Association, Vol. 50, 457-516.

[8]   MONTGOMERY, D.C. (1985): Introduction to Statistical Quality Control. New York: Wiley & Sons.

[9]   SCHILLING, E.G. (1982): Acceptance Sampling in Quality Control. New York: Marcel Dekker.

Design of Economic Sampling Plan Based on Empirical Bayesian Approach

H. Ohta and S. Ogawa, Osaka, Japan

1. INTRODUCTION

It is necessary to estimate the percent nonconforming units in accepted lots for designing an economic sampling plan. Hahn(1986) solves the problem of estimating the number of nonconforming units in lots that pass zero-defect sampling inspection. He shows that the average number of nonconforming units in these lots can be estimated from the number of accepted lots and the number of lots rejected with a single nonconforming unit.

Recently, Zaslavsky(1988) extends Hahn's results. The estimating method given by Zaslavsky is called an Empirical Bayes procedure because the unobserved parameter (number of nonconforming units per lot) of the distribution is regard as a random variable (see Cassella(1955), Martz(1975), Robbins(1955), for example). The procedure requires no assumptions about the production and sampling processes except that the samples are small compared to the lot size and are drawn randomly.

This paper extends Zaslavsky's results in the following directions. Generally, before shipping the accepted lots the nonconforming units found in the samples are removed or replaced by conforming units. Zaslavsky does not consider such a situation. A modification of his method is provided first. It consists in estimating the number of nonconforming units in the lots that pass c -defect sampling inspection. Concerning the determination of optimal sampling plan which minimizes the total expected cost per lot, he gives an assumption that a zero-defect sampling plan is used. A design procedure for the optimal c -defect sampling plan is presented using the multistage decision process techniques and a modified estimating method.

2. ESTIMATION OF THE PERCENT NONCONFORMING UNITS IN ACCEPTED LOTS

Suppose that $T^{(k)}$ lots of each size $N^{(k)}$ were produced in the period k ($k =1,2, \cdots ,K$). Samples of size $n^{(k)}$ are drawn randomly from each lot and

Frontiers in Statistical Quality Control 4
Ed. by Lenz et al.
© Physica-Verlag Heidelberg 1992

inspected. Suppose that in a lot there are $x^{[k]}$ nonconforming units of which $r^{[k]}$ are in the corresponding sample. If $r^{[k]} \leqq c^{[k]}$, then the lot is accepted, otherwise the lot is rejected, where $c^{[k]}$ is the acceptance number of sampling inspection in the period k. Also suppose that $T_i^{[k]}$ is the number of samples which contain exactly i nonconforming units, that is those for which $r^{[k]} = i$.

The number of nonconforming units per lot has some unknown distribution which is determined by the characteristics of the production process; however, we may write $x^{[k]} \sim F$ where F is the unknown distribution. We wish to estimate $E[x_A^{[K]}]$, the expected number of nonconforming units per lot in lots that pass the sampling inspection through K periods. It can be shown that

$$E[x_A^{[K]}] = \sum_k \sum_i \int x^{[k]} P(r^{[k]} = i \mid x^{[k]}, N^{[k]}) dF(x^{[k]})$$

$$\diagup \sum_k \sum_i P(r^{[k]} = i \mid N^{[k]})$$

$$= \sum_k \sum_i (N^{[k]} \diagup n^{[k]})(i+1) P(r^{[k]} = i+1 \mid N^{[k]})$$

$$\diagup \sum_k \sum_i P(r^{[k]} = i \mid N^{[k]}) \tag{1}$$

where $\sum_k$ and $\sum_i$ stand for $\sum_{k=1}^{K}$ and $\sum_{i=0}^{c^{[k]}}$, respectively. By using eqn.(1) the total expected number of nonconforming units which are included in all accepted lots is given by

$$E[X_A^{[K]}] = E[x_A^{[K]}] \times \sum_k \{T^{[k]} \sum_i P(r^{[k]} = i \mid N^{[k]})\}. \tag{2}$$

Generally, before shipping the accepted lots, nonconforming units found in the sample are removed or replaced by conforming units. However, Zaslavsky(1988) does not consider such a situation. Therefore, suppose that nonconforming units found in the samples are replaced by conforming units. The total expected number of nonconforming units which are included in the samples in all accepted lots is given by

$$E[R_A^{[K]}] = \sum_k \sum_i i T_i^{[k]} \tag{3}$$

Furthermore, the total expected number of units which are included in all accepted lots is given by

$$E[U_A^{[K]}] = \sum_k \sum_i N^{[k]} T^{[k]} P(r^{[k]} = i \mid N^{[k]}). \tag{4}$$

By using eqns.(2), (3) and (4), the estimate $\hat{p}_A{}^{[K]}$ of expected percent nonconforming unit in accepted lots through K periods is obtained as

$$\hat{p}_A{}^{[K]} = \{E[X_A{}^{[K]}]-E[R_A{}^{[K]}]\} / E[U_A{}^{[K]}] \ . \tag{5}$$

Replacing the probabilities with the observed proportions of samples with i and $i+1$ nonconforming units, $T_i{}^{[K]} / T^{[K]}$ and, $T_{i+1}{}^{[K]} / T^{[K]}$ respectively, yields

$$\hat{p}_A{}^{[K]} \simeq [\{ \sum_k \sum_i (N^{[K]} / n^{[K]})(i+1)(T_{i+1}{}^{[K]} / T^{[K]})\}\{ \sum_k \sum_i T_i{}^{[K]}\}$$
$$/ \{ \sum_k \sum_i (T_i{}^{[K]} / T^{[K]})\}\{ \sum_k \sum_i N^{[K]} T_i{}^{[K]}\}]$$
$$- \{ \sum_k \sum_i iT_i{}^{[K]} / \sum_k \sum_i N^{[K]} T_i{}^{[K]}\} \tag{6}$$

If $K=1$ and $\sum_k \sum_i iT_i{}^{[K]} = 0$ in eqn.(6), the result corresponds with Zaslavsky's one. In such a manner, the $\hat{p}_A{}^{[K]}$ given by eqn.(6) is characterized by the property of multistage decision processes, and stands for the average outgoing quality AOQ.

3. ACCURACY OF THE ESTIMATE

Table 1 shows the sample data from a zero-defect sam ling plan through 4 periods obtained by a computer simulation under t e conditions that $n^{[1]}=125,$ $n^{[2]}=39,$ $n^{[3]}=98,$ $n^{[4]}=99,$ $N^{[1]}=N^{[2]}=N^{[3]}=N^{[4]}=5000,$ $T^{[1]}=T^{[2]}=T^{[3]}=T^{[4]}=300$ and the average percent noncon rming rate of the production process is 3%. Table 2 shows the estimated va ues $\hat{p}_A{}^{[K]}$ and the true values $p_A{}^{[K]}$ ($K=1,2,3,4$) of the percent nonconformii g rate in accepted lots when $c^{[k]}=0,$ 1 and 2($k=1,2,\cdots,K$). It is evident fro Table 2 that the accuracy of the estimate is getting higher as the accep ance number $c^{[k]}$ and K increase, but the estimate slightly tends to overshoot. This bias is caused by the reason that while $T_i{}^{[K]} / T^{[K]}$ and $T_{i+1}{}^{[K]} / T^{[K]}$ are unbiased estimates of the corresponding probabilities, the ratio $T_{i+1}{}^{[K]} / T_i{}^{[K]}$ is biased upwards (see Zaslavsky(1988)).

Period	\multicolumn{12}{c}{Number of nonconforming units}	Number of lots
	0	1	2	3	4	5	6	7	8	9	10	11	

Period	0	1	2	3	4	5	6	7	8	9	10	11	Number of lots
k=1	6	28	49	64	54	51	29	11	5	1	1	1	300
k=2	96	116	51	37									300
k=3	12	65	58	63	54	20	23	5					300
k=4	13	60	59	57	62	20	25	4					300

Table 1: Distribution of the number of nonconforming units
in lots using a zero-defect sampling plan
($n^{[1]}=125$, $n^{[2]}=39$, $n^{[3]}=98$, $n^{[4]}=99$; $N^{[1]}=N^{[2]}=N^{[3]}=N^{[4]}=5000$;
$T^{[1]}=T^{[2]}=T^{[3]}=T^{[4]}=300$; K=4 periods)

Periods	Acceptance number	Estimated value $\hat{p}_A^{[K]}$	True value $p_A^{[K]}$	Relative error
K=1	$c^{[1]}=0$	0.0373	0.0287	0.30
	$c^{[1]}=1$	0.0295	0.0287	0.03
	$c^{[1]}=2$	0.0303	0.0290	0.04
K=2	$c^{[1]}=c^{[2]}=0$	0.0314	0.0288	0.09
	$c^{[1]}=c^{[2]}=1$	0.0267	0.0297	- 0.10
	$c^{[1]}=c^{[2]}=2$	0.0315	0.0296	0.06
K=3	$c^{[1]}=c^{[2]}=c^{[3]}=0$	0.0339	0.0287	0.18
	$c^{[1]}=c^{[2]}=c^{[3]}=1$	0.0260	0.0290	- 0.10
	$c^{[1]}=c^{[2]}=c^{[3]}=2$	0.0305	0.0293	0.04
K=4	$c^{[1]}=c^{[2]}=c^{[3]}=c^{[4]}=0$	0.0352	0.0286	0.23
	$c^{[1]}=c^{[2]}=c^{[3]}=c^{[4]}=1$	0.0257	0.0279	- 0.08
	$c^{[1]}=c^{[2]}=c^{[3]}=c^{[4]}=2$	0.0296	0.0286	0.03

Table 2: Estimated value $\hat{p}_A^{[K]}$ and true value $p_A^{[K]}$

4. ECONOMIC SAMPLING PLAN

The following cost factors can be defined for designing the economic sampling plan by using the estimate of percent nonconforming units in the accepted lots :

C_r : Cost per unit for replacing the rejected lot (Suppose that the rejected lot is replaced lot by lot),

C_s : Cost per unit for testing units,

C_d : Cost per unit for replacing the nonconforming units contained in the accepted lots(including any costs which arise when an attempt is made to use a nonconforming unit for its intended purpose by consumers).

In what follows we want to decide the sampling plan $(n^{(K+1)}, c^{(K+1)})$ at period $(K+1)$ by using the sample data obtained till period K. In this case, the total expected cost per lot, $E[S^{(K+1)}]$, at period $(K+1)$ would be

$$E[S^{(K+1)}]=C_r N^{(K+1)} \hat{P}_R{}^{(K)} +C_s n^{(K+1)} +C_d N^{(K)} (1- \hat{P}_R{}^{(K)})p_\wedge{}^{(K)}, \qquad (7)$$

where $\hat{P}_R{}^{(K)}$ is the estimate of expected probability of rejecting the lots through K periods, and is given by

$$\hat{P}_R{}^{(K)} =1 - \frac{1}{K} \ \{\sum_k \sum_i P(r^{(k)} =i \mid N^{(k)})\}$$

$$\simeq 1 - \frac{1}{K} \sum_k \sum_i (T_i{}^{(k)} / T^{(k)}) \qquad . \qquad (8)$$

Substitution of eqns.(6) and (8) into eqn.(7) yields

$$E[S^{(K+1)}]=C_r N^{(K+1)} \{1 -\frac{1}{K} \sum_k \sum_i (T_i{}^{(k)} / T^{(k)})\}$$

$$+C_s n^{(K+1)} +C_d N^{(K+1)} [\{ \sum_k \sum_i (N^{(k)} / n^{(k)})(i+1)(T_{i+1}{}^{(k)} / T_i{}^{(k)})\}$$

$$\times \sum_k \sum_i T_i{}^{(k)} - \{ \sum_k \sum_i (T_i{}^{(k)} / T^{(k)})\}\{ \sum_k \sum_i iT_i{}^{(k)} \}]$$

$$/ K \sum_k \sum_i N^{(k)} T_i{}^{(k)} \qquad . \qquad (9)$$

Now, we propose that the sampling plan $(n_{opt}{}^{(K)}, c_{opt}{}^{(K)})$ which minimizes the total expected cost $E[S^{(K+1)}]$ is used for the sampling plan $(n^{(K+1)}, c^{(K+1)})$. Therefore, replacing $n^{(K+1)}$ in eqn.(9) to $n^{(K)}$ and partial differentiating eqn.(9) with respect to $n^{(K)}$, we have

$$n^{(K+1)} =n_{opt}{}^{(K)} =[C_d N^{(K+1)} N^{(K)} \{ \sum_k \sum_i T_i{}^{(k)} \}\{ \sum_i (i+1)(T_{i+1}{}^{(k)} / T^{(k)})\}$$

$$/ C_s K \sum_k \sum_i N^{(k)} T_i{}^{(k)}]^{1/2} \qquad . \qquad (10)$$

An algorithm is as follows which is determining the optimal sample size $n_{opt}^{[K]}$ and acceptance number $c_{opt}^{[K]}$ and which is minimizing the total expected cost $E[S^{[K+1]}]$:

[Step 1] Set C_r, C_s and C_d to be given values and initialize $C^*=\infty$, $n^*=0$ and $c^*=0$.

[Step 2] Set $K=1$ and $c^{[K]}=0$, and $N^{[1]}$ and $N^{[2]}$ to be given values.

[Step 3] Calculate $n^{[K+1]}=n_{opt}^{[K]}$ by eqn.(10) and $E[S^{[K+1]}]$ by eqn.(9). If $E[S^{[K+1]}]<C^*$ then $C^*=E[S^{[K+1]}]$, $n^*=n_{opt}^{[K]}$, $c^*=c^{[K]}$ and go to Step 4. Otherwise, directly go to Step 4.

[Step 4] $c^{[K]}=c^{[K]}+1$. If $c^{[K]} \leqq max\ r^{[K]}$, go to Step 3. Otherwise, go to Step 5.

[Step 5] The current n^* and c^* are the optimal sample size $n_{opt}^{[K]}$ and acceptance number $c_{opt}^{[K]}$ and use them for the sampling plan at period $(K+1)$ as $(n^{[K+1]}, c^{[K+1]})=(n_{opt}^{[K]}, c_{opt}^{[K]})$.

[Step 6] Initialize $C^*=\infty$, $n^*=0$, $c^*=0$ and set $K=K+1$, $c^{[K]}=0$, and $N^{[K]}$ to be given value, then go to Step 3.

Thus, a design procedure for the optimal c -defect sampling plan is obtained using multistage decision process techniques under the condition that the lot size at each period is large enough.

5. NUMERICAL EXAMPLES

For designing the optimal sampling plan at each period by using an empirical Bayesian approach, suppose that samples of size $n^{[1]}=125$ are drawn randomly from $T^{[1]}=300$ lots of each size $N^{[1]}=5000$ and inspected. Table 3 shows the sample data which are same as the data for $K=1$ in Table 1. Table 4 shows the optimal sample size $n_{opt}^{[1]}$ and acceptance number $c_{opt}^{[1]}$ which are dirived from the algorithm above based on the sample data given in Table 3 for various C_r and C_s when $C_d = 1$ and $N^{[2]} = 5000$. As is evident from Table 4, we have following results:

(1) The optimal sample size $n_{opt}^{[1]}$ is monotonously decreasing in the cost C_s per unit for testing units when the cost C_r per unit for replacing the rejected lot, the cost C_d per unit for replacing the nonconforming unit contained in the accepted lots and the optimal acceptance number $c_{opt}^{[1]}$ are constant.

(2) In other words, the optimal sampling plan $(n_{opt}^{[1]}, c_{opt}^{[1]})$ is tightened as the value of the ratio C_d / C_s increases, that is the cost C_d

increases against the cost C_s.

(3) The optimal sampling plan $(n_{opt}^{[1]}, c_{opt}^{[1]})$ is reduced as the cost C_r increases. This is evident from Fig.1 which shows the OC curves of the optimal sampling plans $(n_{opt}^{[1]}, c_{opt}^{[1]}) = (48, 0)$, $(301, 8)$ and $(304, 10)$ in the cases of $C_r = 0.02$, 0.03 and 0.04 (and/or 0.05), respectively when $C_s = 0.2$. Note that in this case the average percent nonconforming rate of the production process is 3%.

Table 5 shows the numerical results of the sampling plan $(n^{[K+1]}, c^{[K+1]}) = (n_{opt}^{[K]}, c_{opt}^{[K]})$ and the minimum total expected cost $E[S^{[K+1]}] (K=1 \sim 4)$ for various C_s when $C_r = 0.03$, $C_d = 1$, $N^{[2]} = 5000$, $N^{[3]} = 8000$, $N^{[4]} = 4000$ and $N^{[5]} = 6000$. For reference, the total expected cost when the sampling plan $(n, c) = (125, 0)$ is used for all periods is also given in Table 5. It is evident from Table 5 that the use of sampling plans obtained by the multistage decision processes is economical and efficient.

Number of non-conforming units	0	1	2	3	4	5	6	7	8	9	10	11
Number of lots	6	28	49	64	54	51	29	11	5	1	1	1

Table 3: Sample data from a zero-defect sampling plan
($n^{[1]} = 125$, $N^{[1]} = 5000$, $T^{[1]} = 300$)

C_s / C_r	0.1	0.2	0.3	0.4
0.02	(426, 8)	(48, 0)	(39, 0)	(34, 0)
0.03	(430, 10)	(301, 8)	(246, 8)	(213, 8)
0.04	(430, 10)	(304, 10)	(246, 8)	(213, 8)
0.05	(430, 10)	(304, 10)	(248, 10)	(214, 9)

Table 4: Optimal sampling plan $(n_{opt}^{[1]}, c_{opt}^{[1]})$
($N^{[1]} = 5000$, $N^{[2]} = 5000$)

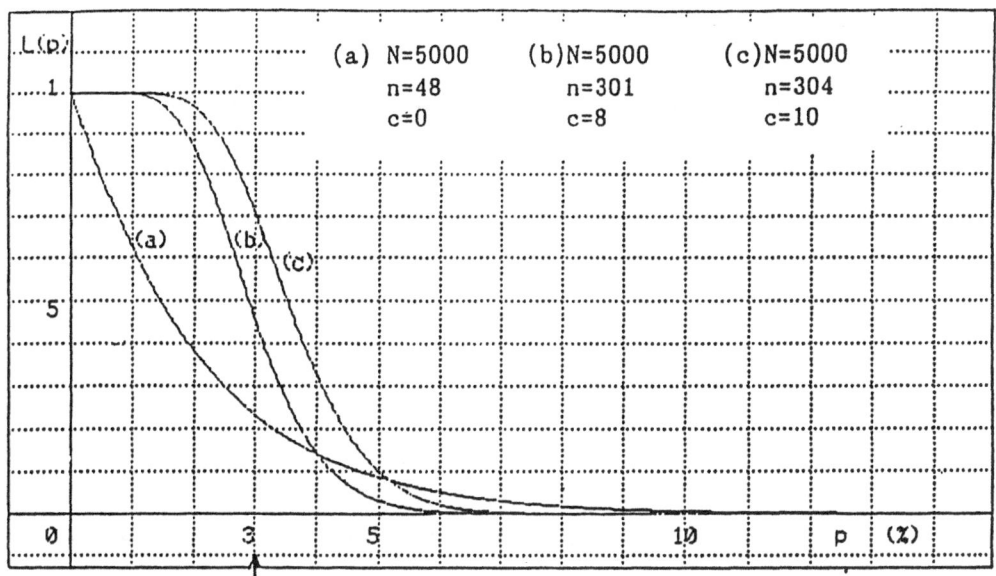

average percent nonconforming unit of the production process

Fig.1: OC-Curves of Sampling Plans (48, 0),(301, 8) and (304,10)

	K=1	K=2	K=3	K=4
$N^{[K+1]}$ Cs	5000	8000	4000	6000
0.1	(430,10) 82.9	(73, 4) 163.1	(88, 4) 75.4	(255, 4) 130.3
	163.2	250.9	131.4	190.8
0.2	(301, 8) 118.4	(78, 2) 194.6	(66, 4) 94.7	(51, 0) 160.7
	175.7	263.4	143.9	203.3
0.3	(246, 8) 145.5	(62, 1) 210.8	(48, 3) 102.5	(116, 3) 160.8
	188.2	275.9	156.4	215.8
0.4	(213, 8) 168.3	(188, 4) 264.9	(9, 0) 108.6	(43, 2) 151.7
	200.7	288.4	168.9	228.3

Table 5: Sampling plan $(n^{[K+1]}, n^{[K+1]})=(n_{opt}^{[K]}, c_{opt}^{[K]})$
and minimum total expected cost
(Cr=0.03)

6. CONCLUDING REMARKS

A modification of Zaslavsky's method based on the empirical Bayesian approach estimating the number of nonconforming units in lots that pass c -defect sampling inspection has been proposed. A design procedure for the optimal c -defect sampling plan which minimizes the total expected cost per lot is also presented using the multistage decision process techniques and a modified estimating method .

7. REFERENCES

[1] Cassella, G. (1985):"An Introduction to Empirical Bayes Data Analysis", *The American Statistician*, Vol.39, No.2, pp.83-87.

[2] Hahn, G. J. (1986):"Estimating the Percent Nonconforming in the Accepted Product After Zero Defect Sampling", *Journal of Quality Technology*, Vol.18, No.3, pp.182-188.

[3] Martz, H. F. (1975):"Empirical Bayes Single Sampling Plans for Specified Posterior Consumer and Producer Risks", *Naval Research Logistics Quarterly*, Vol.22, No.4, pp.661-666.

[4] Robbins, H. (1955):"An Empirical Bayes Approach to Statistics", *Proceedings of Third Berkeley Symposium on Mathematical Statistics and Probability 1*, Berkeley; University of California Press, pp.157-164.

[5] Zaslavsky, A. (1988):"Estimating Nonconformity Rates in c -Defect Sampling", *Journal of Quality Technology*, Vol.20, No.4, pp.248-259.

Minimax Regret Sampling Plans Based on Generalized Moments of the Prior Distribution

W. Seidel, Hamburg, Germany

1. INTRODUCTION

In acceptance sampling, one is often interested in finding sampling plans that incorporate prior information about the fraction of defective items in incoming lots in order to minimize costs. If a prior distribution of the fraction defective is known, a Bayesian sampling plan may be used. Some authors, however, consider this situation to be unlikely and therefore have developed sampling plans based only on incomplete prior information. For example, Krumbholz [6] assumes that a finite set of inequalities

$$\gamma_1^{(i)} \leq \mathbb{P}(p \in I_i) \leq \gamma_2^{(i)}, \quad 1 \leq i \leq n \tag{1}$$

for the fraction defective p is known. Here $I_1 \cup I_2 \cup \ldots \cup I_n$ is a partition of $[0,1]$ into disjoint intervals. A special case of (1) is

$$\mathbb{P}(p \leq a) \geq \gamma. \tag{2}$$

Variables sampling plans incorporating (2) have been developed by Krumbholz and Schröder [7], Schröder [8] and Bruhn-Suhr [1]. v. Collani [3] puts a similar constraint on p, namely

$$\mathbb{P}(p \leq p_0) = \alpha, \tag{3}$$

where p_0 denotes a particular cost parameter. Again variables sampling plans have been developed in the presence of (3): see v. Collani and Unterschemmann [4, 5].

From the viewpoint of the minimax principle, a sampling plan must be chosen that minimizes the maximum average loss that can occur. In the model presented here, calculating the maximum average loss is equivalent to maximizing the integrals of a loss function with respect to the set of prior distributions compatible with the available information. In the papers cited above, this is done by a skilful exploitation of special properties of the particular situations.

We shall present an approach where the prior information is given by a finite number of generalized moment conditions. The calculation of the maximum loss is reduced to an optimization problem in the n–dimensional Euclidean space.

Suppose that the fraction defective p or a related parameter x varies from lot to lot in accordance with a probability distribution π. Usually prior information is gained from past inspections. Doing so, we do not sample from π directly, but from a product πK of π and a transition probability K, as will be shown in section 3. The influence of K has to be considered in some way. Using generalized moments allows us to incorporate properties of πK immediately without being forced to invert the mapping $\pi \mapsto \pi K$.

Frontiers in Statistical Quality Control 4
Ed. by Lenz et al.
© Physica-Verlag Heidelberg 1992

In some cases, πK reduces to a convolution $\pi * v$. Here v is known, but may also vary from lot to lot (see section 3). Thus in the worst case we are given a realization of a random vector $(Z_1,...,Z_n)$, where Z_i is distributed as $\pi * v_i$, $v_i \neq v_j$ for $i \neq j$.

Finally we present an unbiased and consistent estimator for an arbitrary number of moments of π from such samples.

The paper is organized as follows: Section 3 contains the necessary background, section 4 deals with optimization on moment sets and in section 5 we discuss the estimation of the moments of the prior distribution.

2. NOTATION

A Borel measurable subset A of $\mathbb{R}^n$ is always endowed with the Borel σ-algebra $\mathfrak{B}(A)$. Given $A \in \mathfrak{B}(\mathbb{R}^n)$ and $B \in \mathfrak{B}(\mathbb{R}^k)$, a mapping $K : A \times \mathfrak{B}(B) \to [0,1]$ is called a transition probability if it satisfies

i) $K(\cdot,C)$ is measurable for each $C \in \mathfrak{B}(B)$ and
ii) $K(a,\cdot)$ is a probability measure on B for each $a \in A$.

For each probability measure π on A, the product πK defined by

$$\pi K(C) = \int_A K(a,C)\pi(da) \quad (C \in \mathfrak{B}(B))$$

is a probability measure on B.

Given probability measures π on A and v on B, $\pi \otimes v$ denotes the product measure on $A \times B$. If $A = B = \mathbb{R}^n$, $K(a,C) := v(C-a)$ defines a transition probability. For this K, the product πK is the convolution of π and v, written $\pi * v$.

We shall denote the normal distribution with expectation μ and variance σ^2 by $N(\mu,\sigma^2)$, the distribution function of $N(0,1)$ by Φ and its density by φ. δ_x stands for the Dirac measure concentrated at x.

Let $\mathbb{R}_+ := \{x \in \mathbb{R} : x > 0\}$.

3. BACKGROUND FROM QUALITY CONTROL

Let p denote the fraction of defective items in a lot to be inspected. For sampling inspection we restrict our attention to sampling plans (n,c), $n \in \mathbb{N}$ being the sample size and $c \in \mathbb{R}$ an acceptance limit. The lot is accepted if a test statistic, depending on the particular situation, does not exceed c.

We adopt the linear cost model introduced by Stange [9], transformed by the substraction of the "unavoidable loss" and a suitable normalization:

Let $p_0 \in]0,1[$ and $q > 0$. The loss when the lot is accepted is defined as

$$L_a(p,n) := nq + \max \{p-p_0, 0\},$$

whereas the loss when the lot is rejected is

$$L_r(p,n) := nq + \max\{p_0-p, 0\}.$$

The parameter p_0 is called the "break even quality level", q is the (relative) cost of inspecting one item.

Let us assume that a statistical model is given, parametrized by a parameter x that varies in a measurable subset $\mathfrak{X}$ of $\mathbb{R}^k$. To each $x \in \mathfrak{X}$ corresponds uniquely a fraction defective p(x), and a unique probability A(x;n,c) of accepting the lot is associated to every sampling plan (n,c) and each $x \in \mathfrak{X}$.

We give three examples:

Example 1 Normally distributed quality characteristic with known variance σ^2 and unknown expectation μ.

Here we may choose $\mathfrak{X} = \mathbb{R}$, $\mathfrak{X} \ni x = \mu$ the expectation. Let $\overline{X}_n$ denote the mean of a sample of size n.

Testing against double specification limits L < U yields

$$p(\mu) = \Phi((\mu-U)/\sigma) + \Phi((L-\mu)/\sigma).$$

Define $\mu_0 := (L+U)/2$ and assume that a lot is accepted iff

$$|(\overline{X}_n-\mu_0)/\sigma| \leq c \qquad (c > 0).$$

Then

$$A(\mu;n,c) = \Phi\left(\sqrt{n}(c+(\mu_0-\mu)/\sigma)\right) - \Phi\left(\sqrt{n}(-c+(\mu_0-\mu)/\sigma)\right).$$

Example 2 $N(\mu,\sigma^2)$–distributed qualtity characteristic, both μ and σ^2 being unknown.

A natural choice of $\mathfrak{X}$ is $\mathfrak{X} = \mathbb{R} \times \mathbb{R}_+$, $\mathfrak{X} \ni x = (\mu,\sigma^2)$. Given an upper specification limit U, we obtain

$$p(\mu,\sigma^2) = \Phi((\mu-U)/\sigma);$$

the formula for $p(\mu,\sigma^2)$ in the two–sided case is the same as in example 1.

Given a random sample $(X_1,...,X_n)$ from the parent distribution, let

$$S_n^2 := (n-1)^{-1}\sum_{i=1}^{n}(X_i-\overline{X}_n)^2.$$

In the one–sided case, the usual sampling plan is based on the test statistic $(\overline{X}_n,S_n^2)$. In the two–sided case, there are several possibilities for constructing sampling plans, c.f. Bruhn–Suhr/Krumbholz [2]. Again, the efficient ones start from $(\overline{X}_n,S_n^2)$.

Example 3 Sampling inspection by attributes.

Given a lot of size N, let $\mathfrak{X} = \{0,1,...,N\}$, $x \in \mathfrak{X}$ the number of defective items in the lot. If we draw a random sample of size n with (without) replacement, the number D of defective items in the sample has a binomial (hypergeometric) distribution. The lot is accepted, iff D does not exceed an acceptance number c.

Given $x \in \mathfrak{X}$, using a sampling plan (n,c) results in an average loss

$$R(x;n,c) = L_a(p(x),n)A(x;n,c) + L_r(p(x),n)(1-A(x;n,c)).$$

In leaving (n,c) fixed, the function

$$R(\cdot;n,c) : \mathfrak{X} \to [0,1]$$

is called a "regret function". Here we may always assume that R is measurable. Therefore, given a probability measure π on $\mathfrak{X}$, it is possible to define

$$r(\pi;n,c) := \int_{\mathfrak{X}} R(x;n,c)\pi(dx),$$

the "Bayes risk" of (n,c), given π (π considered as a prior distribution of x).

Denote by M the set of all probability measures on $\mathfrak{X}$. Any subset Π of M describes prior information about x. A sampling plan (n_Π,c_Π) is called a "Π-minimax regret sampling plan" if it satisfies

$$\sup_{\pi \in \Pi} r(\pi;n_\Pi,c_\Pi) = \inf_{n,c} \sup_{\pi \in \Pi} r(\pi;n,c).$$

For example, $\Pi = \{\pi\}$ yields a Bayesian sampling plan, whereas $\Pi = M$ results in the usual minimax regret sampling plan (without prior knowledge about x). Write $\mathfrak{X}_a := p^{-1}([0,a])$, then inequality (2) corresponds to

$$\Pi = \{\pi \in M : \pi(\mathfrak{X}_a) \geq \gamma\},$$

whereas (3) corresponds to

$$\Pi = \{\pi \in M : \pi(\mathfrak{X}_{p_0}) = \alpha\}.$$

Further examples will be given in section 4.

In typical applications, prior information is obtained from records of past inspections. With each sampling plan considered above, there is, in a natural way, an estimator Z of x associated, which is a random variable with values in $\mathfrak{X}$. Let us denote its distribution, if x is the true parameter, by $K(x,\cdot)$ (by $K^{(n)}(x,\cdot)$, if the sample size is mentioned explicitly). In each of the above examples, $K : \mathfrak{X} \times \mathfrak{B}(\mathfrak{X}) \to [0,1]$ is a transition probability:

Example 1 (continued)   Here $Z = \overline{X}_n$ and $K^{(n)}(\mu,\cdot) = N(\mu,\sigma^2/n)$.

Example 2 (continued) $Z = (\overline{X}_n,S_n^2)$ and $K^{(n)}((\mu,\sigma^2),\cdot) = N(\mu,\sigma^2/n) \otimes v(n-1,\sigma^2)$, where $v(k,a)$ is the distribution of $(a/k)Y$, Y having a chi-square distribution with k degrees of freedom.

Example 3 (continued)   Here Z is the number of defective items in the sample, and $K(x,\cdot)$ is the binomial or hypergeometric distribution corresponding to the fraction defective x/N.

If the parameter x has a probability distribution π, the observed quantity Z is distributed as πK. Now assume, that we are inspecting a sequence of lots, where x varies independently from lot to lot according to π. Assume further that we are inspecting the i-th lot using a sampling plan (n_i, c_i). We allow for $n_i \neq n_j$, $i \neq j$, for instance, because we may switch from time to time to an updated sampling plan which incorporates the prior information gained so far. In this case, we observe a sequence $(Z_i)_{i \in \mathbb{N}}$ of independent random variables, where Z_i is distributed as $\pi K^{(n_i)}$. The task is to extract prior information which can be used to calculate a minimax regret sampling plan.

Sometimes two simplifications are possible:

S1) If $n_i = n$, $i \in \mathbb{N}$, we are given a sample from one particular distribution $\pi K^{(n)}$.

S2) $\mathfrak{X} = \mathbb{R}^k$ and $K^{(n)}(x,C) = v^{(n)}(C-x)$, $C \in \mathfrak{B}(\mathfrak{X})$ and $v^{(n)}$ some probability distribution on $\mathfrak{X}$ (see example 1: Here $k = 1$ and $v^{(n)} = N(0, \sigma^2/n)$). Then Z_i is distributed as $\pi K^{(n_i)} = \pi * v^{(n_i)}$. For such convolutions, estimating properties of π is simpler than for arbitrary transition probabilities.

Prior information suitable for practical purposes should meet two requirements:

R1) It should be easily obtainable from past inspection records, and

R2) maximization of the Bayes risk on the corresponding set Π should be possible.

In the approach presented in section 4, Π is given by a finite set of generalized moment conditions. In this case R2 reduces to a constrained optimization problem in n real variables (which of course may still be hard to solve). Now consider R1: For S1 it will be possible to incorporate conditions on πK directly without having to invert $\pi \mapsto \pi K$. Under the conditions of S2, an unbiased and consistent estimator of the moments of π is developed in section 5 for $k = 1$ (it may be generalized to arbitrary k).

4. OPTIMIZATION ON MOMENT SETS

Let $\mathfrak{X} \in \mathfrak{B}(\mathbb{R}^k)$ and let us denote the set of all probability measures on $\mathfrak{X}$ by M. A set $\Pi \subseteq M$ is called <u>convex</u>, iff $\pi_1, \pi_2 \in \Pi$ and $\alpha \in]0,1[$ imply $\pi = \alpha\pi_1 + (1-\alpha)\pi_2 \in \Pi$. If Π is convex then $\pi \in \Pi$ is called an <u>extreme point</u> of Π if the last formula implies $\pi_1 = \pi_2$. The set of extreme points of Π is denoted by ex Π.

Theorem 1 For each $i = 1,...,n$, let $f_i : \mathfrak{X} \to \mathbb{R}$ be a measurable function and $I_i \subseteq \mathbb{R}$ a (possibly degenerate) closed interval. Then the set

$$\Pi = \left\{ \pi \in M: f_i \text{ is } \pi\text{-integrable and } \int_{\mathfrak{X}} f_i \, d\pi \in I_i, \, 1 \leq i \leq n \right\}$$

is convex. Furthermore, the following statements hold:

a) $\text{ex } \Pi \subseteq \Big\{ \pi \in \Pi : \pi = \sum_{i=1}^{m} t_i \, \delta_{x_i}, \ t_i > 0, \ \sum_{i=1}^{m} t_i = 1, \ x_i \in \mathfrak{X}, \ 1 \le m \le n+1,$

the vectors $(1, f_1(x_i),...,f_n(x_i))$, $1 \le i \le m$, are linearly independent $\Big\}$.

b) Let $g : A \to \mathbb{R}$ be integrable for every $\pi \in \Pi$ (possibly with integral values ∞ or $-\infty$). Then

$$\sup\Big\{ \int_{\mathfrak{X}} g \, d\pi : \pi \in \Pi \Big\} = \sup\Big\{ \int_{\mathfrak{X}} g \, d\pi : \pi \in \text{ex } \Pi \Big\}.$$

A set Π defined as in theorem 1 will be called a "moment set", each of the conditions $\int f_i \, d\pi \in I_i$ a "generalized moment condition". For $\mathfrak{X} = \mathbb{R}$ and $I_i =]-\infty,0]$, $1 \le i \le n$, part a) and b) of theorem 1 correspond to proposition 3 and 2 in Winkler [10]. To get constraints of the form $\int f_i \, d\pi \in [a_i,b_i]$, let a_i-f_i and f_i-b_i be in the moment set. The condition of linear independency in proposition 3 guarantees that the extremal measures are still concentrated at a number of points not exceeding $n+1$. The proof of proposition 3 and 2 in a more general setting may be found in Winkler [11].

Corollary   Denote by $R: \mathfrak{X} \to [0,1]$, $x \mapsto R(x)$ the regret function we are interested in (omitting the variables n and c), and let Π be given as in theorem 1. Then $\sup\limits_{\pi \in \Pi} \int_{\mathfrak{X}} R(x)\pi(dx)$ is the solution to the following problem:

$$\text{Maximize} \atop t_j, x_j \quad \sum_{j=1}^{n+1} t_j \, R(x_j)$$

subject to $x_j \in \mathfrak{X}, \ 1 \le j \le n+1$

$t_j \ge 0, \ 1 \le j \le n+1$

$\sum_{j=1}^{n+1} t_j = 1,$

$\sum_{j=1}^{n+1} t_j \, f_i(x_j) \in I_i, \quad 1 \le i \le n.$

Note, that simplifications may be possible if some vectors $(1, f_1(x_i),...,f_n(x_i))$ turn out to be linearly dependent.

Some examples illustrate the approach:

Example 4 $\mathfrak{X} = \mathbb{R}$ and $f_i(x) = x^i$, $i = 1,...,n$. Then $\int_{\mathfrak{X}} f_i \, d\pi$ is the i-th moment of π. $I_i = \{c_i\}$ means that we assume the i-th moment of the prior to be known exactly, whereas $I_i = [l_i,u_i]$ $(l_i < u_i)$ may be a confidence interval for this moment.

Example 5   Given $A \in \mathfrak{B}(X)$, let $f := 1_A$, the characteristic function of A.

$$\Pi = \Big\{ \pi \in M : \int_{\mathfrak{X}} f \, d\pi \in I \Big\},$$

I a closed interval, corresponds to the prior information

$$\mathbb{P}(x \in A) \in I,$$

which covers among others (2) and (3). Here

$$\text{ex } \Pi \subseteq \left\{ \pi = \alpha \, \delta_x + (1-\alpha)\delta_y : x \in A, \, y \notin A, \, \alpha \in I \cap [0,1] \right\}.$$

Let us now consider moment conditions on the distribution πK of an observable estimator of the unknown parameter x:

Let $K : \mathfrak{X} \times \mathfrak{B}(\mathfrak{X}) \to [0,1]$ be a transition probability and $f_i : \mathfrak{X} \to \mathbb{R}$ be integrable with respect to $K(x,\cdot)$ for each $x \in \mathfrak{X}$. For closed intervals $I_1,...,I_n$ consider

$$\Pi = \left\{ \pi \in M : f_i \text{ is } \pi K\text{-integrable and } \int_{\mathfrak{X}} f_i \, d(\pi K) \in I_i, \, i = 1,...,n \right\}.$$

For each $i \in \left\{ 1,...,n \right\}$, define $g_i := Kf_i$ by

$$g_i(x) := Kf_i(x) := \int_{\mathfrak{X}} f_i(y)K(x,dy).$$

g_i is measurable, and if f_i is πK-integrable, then g_i is π-integrable and satisfies

$$\int_{\mathfrak{X}} g_i(x)\pi(dx) = \int_{\mathfrak{X}} f_i(x)(\pi K)(dx).$$

Therefore $\Pi \subseteq \Pi^*$ holds for the set Π^* defined by

$$\Pi^* := \left\{ \pi \in M : g_i \text{ is } \pi\text{-integrable and } \int_{\mathfrak{X}} g_i \, d\pi \in I_i, \, i = 1,...,n \right\}.$$

On the other hand, $\text{ex } \Pi^*$ is a subset of Π, since every $v \in \text{ex } \Pi^*$ is a linear combination of Dirac measures and f_i was assumed to be $K(x,\cdot)$-integrable for each $i \in \left\{ 1,...,n \right\}$ and $x \in \mathfrak{X}$. Thus we obtain for any function g on $\mathfrak{X}$, which is integrable for every $\pi \in \Pi^*$:

$$\sup\left\{ \int_{\mathfrak{X}} g \, d\pi : \pi \in \Pi \right\} = \sup\left\{ \int_{\mathfrak{X}} g \, d\pi : \pi \in \text{ex } \Pi^* \right\}.$$

Let us consider two special cases:

Example 6   Let $A \in \mathfrak{B}(\mathfrak{X})$, $I \subseteq \mathbb{R}$ a closed interval and

$$\Pi = \left\{ \pi \in M : (\pi K)(A) \in I \right\} = \left\{ \pi \in M : \int_{\mathfrak{X}} f \, d\pi \in I \right\}$$

with $f(\cdot) = K(\cdot,A)$. Π corresponds to the prior information

$$\mathbb{P}(Z \in A) \in I, \tag{4}$$

Z being the observable estimator discussed in section 3.

If $K(x,B) = v(B-x)$, v a probability measure on $\mathfrak{X} = \mathbb{R}$, $A = \,]-\infty,a]$ and $I = [\gamma,\infty[$, (4) reads as

$$\mathbb{P}(Z \le a) \ge \gamma, \tag{5}$$

which is similar to (2), but easier to estimate: For instance, if a is the acceptance limit and Z the test statistic, then γ may be estimated by the fraction of accepted lots in past inspections (possibly with a correction that protects against underestimating).

Maximizing the integral of a regret function R subject to (5) is equivalent to the problem

$$\text{Maximize} \quad t\,R(x) + (1-t)R(y)$$
$$\substack{t,x,y}$$

$$\text{subject to} \quad x,y \in \mathbb{R}$$
$$0 \le t \le 1$$
$$t\,v(]-\infty,a-x]) + (1-t)v(]-\infty,a-y]) \ge \gamma.$$

Example 7   Consider again $\mathfrak{X} = \mathbb{R}$ and let $f_i(x) = x^i$, $i = 0,...,n$. For $\mu \in M$ define

$$m_i(\mu) := \int_{\mathbb{R}} f_i \, d\mu$$

(if it exists), the i-th moment of μ. Assume that information on $m_i(\pi K)$, $i = 1,...,n$, has been gained by past inspections. The functions g_i which translate the information on $m_i(\pi K)$ into moment conditions on π admit a simple interpretation:

$$g_i(x) = Kf_i(x) = \int_{\mathfrak{X}} y^i \, K(x,dy)$$

is the i-th moment of $K(x,\cdot)$.

If there exists $v \in M$ such that $K(x,B) = v(B-x)$ and if $m_n(v)$ is finite, then $g_k(x)$ exists for each $k \le n$ and $x \in \mathfrak{X}$ and may be obtained as follows:

$$g_k(x) = \int_{\mathbb{R}} y^k \, K(x,dy) = \int_{\mathbb{R}} (x+y)^k v(dy) = \sum_{i=0}^{k} \binom{k}{i} x^{k-i} \left(\int_{\mathbb{R}} y^i v(dy) \right)$$
$$= \sum_{i=0}^{k} \binom{k}{i} x^{k-i} m_i(v).$$

5. ESTIMATION OF MOMENTS

Let $(X_n)_{n \in \mathbb{N}}$ and $(Y_n)_{n \in \mathbb{N}}$ be two independent sequences of independent random variables on some probability space $(\Omega, \mathfrak{A}, \mathbb{P})$. The expectation of a random variable Z is denoted by $E(Z)$. Assume that the variables X_n are identically distributed with the generic variable X, and that for some $N \in \mathbb{N}$ both $E(|X^k|)$ and $E(|Y_n^k|)$ are finite for $k = 1,...,N$ and each $n \in \mathbb{N}$. Define $m_k := E(X^k)$, $k = 0,...,N$, and $Z_n := X_n + Y_n$, $n \in \mathbb{N}$.

Now let us consider the following situation: $(Z_n)_{n \in \mathbb{N}}$ is observable, whereas $(X_n)_{n \in \mathbb{N}}$ and $(Y_n)_{n \in \mathbb{N}}$ are not; $E(Y_n^k)$ is known for each $k = 1,...,N$, $n \in \mathbb{N}$, whereas m_k is not and therefore has to be estimated.

If all the considered moments exist,

$$E(Z_n^k) = E((X_n+Y_n)^k) = \sum_{j=0}^{k} \binom{k}{j} E(X_n^j)E(Y_n^{k-j}) = m_k + \sum_{j=0}^{k-1} \binom{k}{j} m_j \, E(Y_n^{k-j})$$

holds, from which the equation

$$m_k = E(Z_n^k) - \sum_{j=0}^{k-1} \binom{k}{j} m_j \, E(Y_n^{k-j}) \tag{6}$$

follows. Eq. (6) suggests to define for each $n \in \mathbb{N}$ estimators $M_1^{(n)},...,M_N^{(n)}$ of $m_1,...,m_N$ inductively, based on the sample $(Z_1,...,Z_n)$, as

$$M_k^{(n)} := \frac{1}{n} \sum_{i=1}^{n} \left(Z_i^k - \sum_{j=0}^{k-1} \binom{k}{j} M_j^{(n)} E(Y_i^{k-j}) \right).$$

Observe that $M_0^{(n)} = 1$ for each n, so the recursion is well-defined.

The properties of these estimators are given by

Theorem 2

a) For each $n \in \mathbb{N}$ and $k = 1,...,N$, $M_k^{(n)}$ is an unbiased estimator of m_k.

b) If in addition $E(|X^k|)$ exists for all $k = N+1,...,2N$ and the sequence $(E(|Y_n^k|))_{n \in \mathbb{N}}$ is bounded for each $k = 1,...,2N$, then the sequence $(M_k^{(n)})_{n \in \mathbb{N}}$ is consistent for m_k for each $k = 1,...,N$.

The proof is given in the appendix.

Example 8 For each $i \in \mathbb{N}$ let $n_i \in \mathbb{N}$ and $Y_i \sim N(0,\sigma^2/n_i)$. Suppose that we want to estimate the first three moments of X from $(Z_1,...,Z_n)$. As $E(Y_i) = E(Y_i^3) = 0$ and $E(Y_i^2) = \sigma^2/n_i$, we obtain

$$M_1^{(n)} = \frac{1}{n} \sum_{i=1}^{n} Z_i =: \overline{Z}_n,$$

$$M_2^{(n)} = \frac{1}{n} \sum_{i=1}^{n} (Z_i^2 - \sigma^2/n_i),$$

$$M_3^{(n)} = \frac{1}{n} \sum_{i=1}^{n} (Z_i^3 - 3\overline{Z}_n(\sigma^2/n_i)).$$

6. APPENDIX

Proof of theorem 2

It follows from the assumptions of section 5 that $E(Z_n^k)$ exists for $k = 1,...,N$ ($k = 1,...,2N$ in (b)).

a) is immediately proved by induction on k using (6).

b) From the additional assumptions it follows that

$$\sum_{n \in \mathbb{N}} \frac{\text{Var}(Z_n^k)}{n^2} < \infty$$

holds for each $k = 1,...,N$. Thus we obtain from the law of large numbers, Kolmogorov's criterion,

$$\frac{1}{n} \sum_{i=1}^{n} (Z_i^k - E(Z_i^k)) \xrightarrow{a.s.} 0 \text{ as } n \to \infty$$

for each $k \in \{1,...,N\}$. From (6) it follows that

$$\frac{1}{n} \sum_{i=1}^{n} Z_i^k - \sum_{j=0}^{k-1} \binom{k}{j} m_j \left(\frac{1}{n} \sum_{i=1}^{n} E(Y_i^{k-j}) \right) \xrightarrow{\text{a.s.}} m_k \quad (n \to \infty). \tag{7}$$

Let $\Omega' \in \mathfrak{A}$ such that (7) holds on Ω' for each $k = 1, \ldots, N$, then $\mathbb{P}(\Omega') = 1$. Keeping $\omega \in \Omega'$ fixed, we shall prove

$$M_k^{(n)}(\omega) \to m_k \quad (n \to \infty)$$

by induction on k.

The assertion is true for $k = 1$. In order to prove it for $k > 1$, define

$$a_n := \frac{1}{n} \sum_{i=1}^{n} (Z_i(\omega))^k ,$$

$$b_j^n := \binom{k}{j} \left(\frac{1}{n} \sum_{i=1}^{n} E(Y_i^{k-j}) \right) \quad \text{and}$$

$$m_k^n := M_k^{(n)}(\omega).$$

With these abbreviations, (7) reads as

$$a_n + \sum_{j=0}^{k-1} m_j b_j^n \to m_k \quad (n \to \infty) \tag{8}$$

and the inductive hypothesis is

$$m_j^n \to m_j \quad (n \to \infty), \quad j = 0,1,\ldots,k-1. \tag{9}$$

We have to show $|m_k^n - m_k| \to 0$ $(n \to \infty)$. But

$$|m_k^n - m_k| = |a_n + \sum_{j=0}^{k-1} m_j^n b_j^n - m_k|$$

$$= |a_n + \sum_{j=0}^{k-1} (m_j^n - m_j) b_j^n + \sum_{j=0}^{k-1} m_j b_j^n - m_k|$$

$$\leq |a_n + \sum_{j=0}^{k-1} m_j b_j^n - m_k| + |\sum_{j=0}^{k-1} (m_j^n - m_j) b_j^n|$$

$$\leq |a_n + \sum_{j=0}^{k-1} m_j b_j^n - m_k| + \sum_{j=0}^{k-1} |m_j^n - m_j| |b_j^n|.$$

By (8), the left summand converges to zero, and by (9), each $|m_j^n - m_j|$, $j = 0, \ldots, k-1$, converges to zero. So we may complete the proof by showing that for each $j = 0, \ldots, k-1$, the sequence $(|b_j^n|)_{n \in \mathbb{N}}$ is bounded. But according to the assumption, there exists for each $j = 0, \ldots, 2N$ a constant B_j such that $E(|Y_i^j|) \leq B_j$ for every $i \in \mathbb{N}$. So

$$|b_j^n| = \binom{k}{j} \frac{1}{n} |\sum_{i=1}^{n} E(Y_i^{k-j})|$$

$$\leq \binom{k}{j} \frac{1}{n} \sum_{i=1}^{n} E(|Y_i^{k-j}|)$$

$$\leq \binom{k}{j} \frac{1}{n} n\, B_{k-j} = \binom{k}{j} B_{k-j}.$$

This completes the proof.

7. REFERENCES

[1] BRUHN-SUHR, M. (1988): Kostenoptimale Variablenprüfpläne für den Fall der Normal-
     verteilung mit unbekannter Varianz.
     Dissertation, Universität der Bundeswehr Hamburg.

[2] BRUHN-SUHR, M. and KRUMBHOLZ, W. (1990): A new variables sampling plan for normally
     distributed lots with unknown standard deviation and double
     specification limits,
     Statistical Papers, Vol. 31, 195-207.

[3] COLLANI, E. v. (1986): The α-optimal sampling scheme,
     Journal of Quality Technology, Vol. 18, 63-66.

[4] COLLANI, E. v. and UNTERSCHEMMANN, H. (1989): Alpha-optimal sampling plans for
     variables in the one-sided case.
     Institut für Angewandte Mathematik und Statistik, Würzburg.

[5] COLLANI, E. v. and UNTERSCHEMMANN, H. (1989): Alpha-optimal sampling plans for
     variables in the two-sided case.
     Institut für Angewandte Mathematik und Statistik, Würzburg.

[6] KRUMBHOLZ, W. (1982): Die Bestimmung einfacher Attributprüfpläne unter Berücksich-
     tigung von unvollständiger Vorinformation,
     Allgemeines Stat. Archiv, Vol. 66, 240-253.

[7] KRUMBHOLZ, W. and SCHRÖDER, J. (1987): Zur Ausnutzung unvollständiger Vorinfor-
     mation bei der Minimax-Regret-Methode,
     Allgemeines Stat. Archiv, Vol. 71, 117-125.

[8] SCHRÖDER, J. (1987): Möglichkeiten der Minimax-Regret-Methode bei der messenden
     Prüfung.
     Dissertation, Universität der Bundeswehr Hamburg.

[9] STANGE, K. (1964): Die Berechnung wirtschaftlicher Pläne für die messende Prüfung,
     Metrika, Vol. 8, 48-82.

[10] WINKLER, G. (1982): Integral representation and upper bounds for stop-loss premiums
      under constraints given by inequalities,
      Scand. Actuarial J., 15-21.

[11] WINKLER, G. (1988): Extreme points of moment sets,
      Mathematics of Operations Research, Vol. 13, 581-587.

Part 2
Process Quality Control

Shewhart Control Charts for Individuals with Time-Ordered Data

M. K. Hart and R. F. Hart, Oshkosh, Wisconsin, USA

1. INTRODUCTION: THE RUN CHART AND CONTROL CHARTS FOR TIME-ORDERED DATA

The analysis of time-ordered data is of particular interest to the quality engineer and control statistician to whom the time-series plot is a known as a "run chart". Deming [1982 pp. 112-114 and 1986 pp. 6-7, 310-314] has been a strong proponent of the use of run charts for the analysis of process data. It has been the authors' experience that many of the improvements that are made in industrial processes by using "statistical process control" (SPC) methods result from the use of simple run charts.

It is a logical step to proceed from a run chart on individuals to a Shewhart control chart for individuals. The familiar X and MR chart has become quite widely used for this purpose, but is subject to several difficulties, both in the economics of application and in interpretation. This paper offers two modifications to the X and MR chart which overcome most of its shortcomings, showing examples of each.

2. HOMOGENEOUS SUBGROUPS: THE FUNDAMENTAL ASSUMPTION

The use of homogeneous subgroups of data is the fundamental assumption in the control statistics of Walter Shewhart. The importance of homogeneous subgroups derives from the fact that it is the within-subgroup estimate of σ that is used to set the control chart limits. Homogeneous subgroups have little within-subgroup variation leading to tight, and hence effective, control limits. Lack of homogeneity within the subgroup will result in an inflated estimate of σ and hence inflated control limits. This may potentially render the control chart useless, since assignable causes of variation may be masked by the inflated control chart limits. As Shewhart noted, small time-ordered subgroups of consecutive pieces tend to be homogeneous. The subgroup size for time-ordered data is the central topic for this paper.

Frontiers in Statistical Quality Control 4
Ed. by Lenz et al.
© Physica-Verlag Heidelberg 1992

3. TRADITIONAL SHEWHART CONTROL CHARTS FOR TIME-ORDERED DATA

When seeking evidence of lack of statistical uniformity over time, Shewhart took 100 time-ordered observations, using 25 subgroups of size four, and then stated [1931 pp. 313-314]:

> ...there is nothing sacred about the number four...obviously, if the cause system is changing, the sample size should be as small as possible so that averages of samples do not mask the change. In fact single observations would be the most sensitive to such changes. Why then do we not use a sample size of unity? The answer is that if we do, we are faced with the difficulty of choosing the standard deviation to be used in the control charts...

The primary requirement for a subgroup is that it be as homogeneous as possible. If subgroups are made larger than necessary, or if the subgroup is not comprised of consecutive pieces in time order, this requirement for homogeneity will be unnecessarily violated, tending to render the control chart impotent due to inflated control limits.

Following Shewhart, control charts for time-ordered data in recent years have often used a subgroup size of four. A subgroup size of five has also been held in considerable favor, the logic apparently being "the bigger the better". Clearly this increase in subgroup size for time-ordered data has been a step in the wrong direction, apparently due to lack of understanding for the need for homogeneity within the subgroup and of how that need may be satisfied.

The problem of estimating σ for a subgroup size of unity has been resolved since Shewhart's 1931 work. The most common method has been through the use of the "moving range" of subgroups of size two. Following Western Electric [1956 pp. 21-23] the acronym MR will be used here for the moving range, although some authorities use R, which fails to discriminate between the independent ranges of the usual $\bar{X}$ and R chart and the non-independent moving ranges [ASTM 1976 pp. 99, 130-133; Grant and Leavenworth 1980 pp. 296-298]. The benefits and some shortcomings of the X and MR chart will be discussed in the next section, after which some helpful modifications to this chart will be introduced.

4. THE X AND MR CHART FOR TIME-ORDERED DATA

Deming has called the Shewhart control charts "simple but powerful tools". In this context the X and MR chart would have to be called very simple and very powerful. The power of the chart for individuals was predicted by Shewhart and has been verified in the many applications where the X and MR chart has been used for process improvement. The simplicity is even greater than might have been anticipated.

The X and MR chart is really only an extension of the run chart; the operator quickly becomes comfortable with it and assumes ownership of it. A tremendous simplification results from the fact that the control limits on X and the "natural" ($\pm 3\sigma$) limits of the individuals are one and the same and may be compared directly with the specification limits. It is difficult to overstate the amount of confusion that results in practice because the control limits on an $\bar{X}$ chart do not show the dispersion of the individuals and must not be compared to the specification limits.

The calculation of the moving ranges and the control limits for the X and MR chart are simpler than for $\bar{X}$ and R charts, since the factors are always the same and do not have to be looked up in a table. The estimate of σ that Shewhart found elusive is made just as it would be from a conventional range chart (R chart) with a subgroup size of two.

For all of its simplicity and power, the X and MR chart suffers from some major difficulties. By the very nature of this chart for individuals and moving ranges, a long uninterrupted run of product is needed in order to obtain a sufficient number of observations to make a useful chart. However, today the trend is more and more toward short runs of product in order to hold work-in-process inventories to a minimum.

A second problem with the X and MR chart is that the consecutive pieces of product needed obtain homogeneity of the moving subgroups results in a requirement for 100 percent inspection. Note that if the sampling were to be done by taking an individual observation now, and another at some later time, inflation of the estimate of σ would be assured. This has nothing to do with the type of control chart or classical statistical method used to estimate σ. There is strong motivation to eliminate 100 percent inspection in most processes, which makes the use of the (ordinary) X and MR chart undesirable.

A third difficulty with the X and MR chart arises out of the fact that successive moving ranges are not independent. As a result, the usual criteria for evidence of lack of statistical uniformity are not applicable to the MR chart. For example, a run of eight moving ranges below the centerline does not constitute evidence of lack of uniformity, nor does two out of three moving ranges outside of the two σ upper control limit. Each of these events would be viewed as

evidence of lack of uniformity on any other control chart in accordance
with the commonly used criteria [Western Electric 1956 pp. 26-27; Hart
and Hart 1989 pp. 144-147].

The modifications to the X and MR chart described in the following
sections have proven to be effective in overcoming the three
shortcomings of the X and MR chart discussed above.

5. THE "INTERRUPTED" X AND MR CHART FOR TIME-ORDERED DATA

Consider the first X and MR chart difficulty listed in the
preceding section: its inability to accommodate the common occurrence
of short runs of product, between which there may be tool adjustment or
set-up changes. This problem is easily overcome by interrupting the X
and MR chart at the same places where the process is interrupted. The
resulting chart may be called an "interrupted" or "discontinuous" X and
MR chart [Hart and Hart 1989 pp. 255-258]. The following short data
set of time-ordered X values will be used to illustrate how the X and
MR chart may be modified for an interrupted process. The slash marks
indicate process interruptions.
13, 15, 15, 14 / 6, 7, 2, 3 / 12 / 10, 11 / 13, 11, 8, 12, 14 / 4, 1,
5, 4

An ordinary X and MR chart for this data set, made as if the
process had not been interrupted, is shown in Figure 1a. The letters U
and L denote the 3σ control limits and the 2σ limits are dotted.

The X and MR chart in Figure 1a fails to reveal the lack of
statistical uniformity. This is because the process interruptions were
not taken into account, resulting in inflated MR values, an inflated
$\overline{MR}$, and inflated control limits.. The interruptions to the process can
properly be taken into account if the moving ranges immediately after
these interruptions are either simply not calculated at all or are
given a special status in two regards. First, they are not included in
the calculation of the centerline for the moving range chart, $\overline{MR}$.
Second, each such special moving range should be clearly identified for
interpretation of the chart. They are ranges that occur between runs
of product, not within runs. As such, these moving ranges may often
have higher values than the within-run values. Using them to calculate
control limits would result in inflated control limits, weakening the
control chart. The "Interrupted X and MR chart" for the same data set
is shown in Figure 1b. In Figure 1b, the # symbol denotes a point
outside of the three σ limits and < or > denotes a point off scale.

127

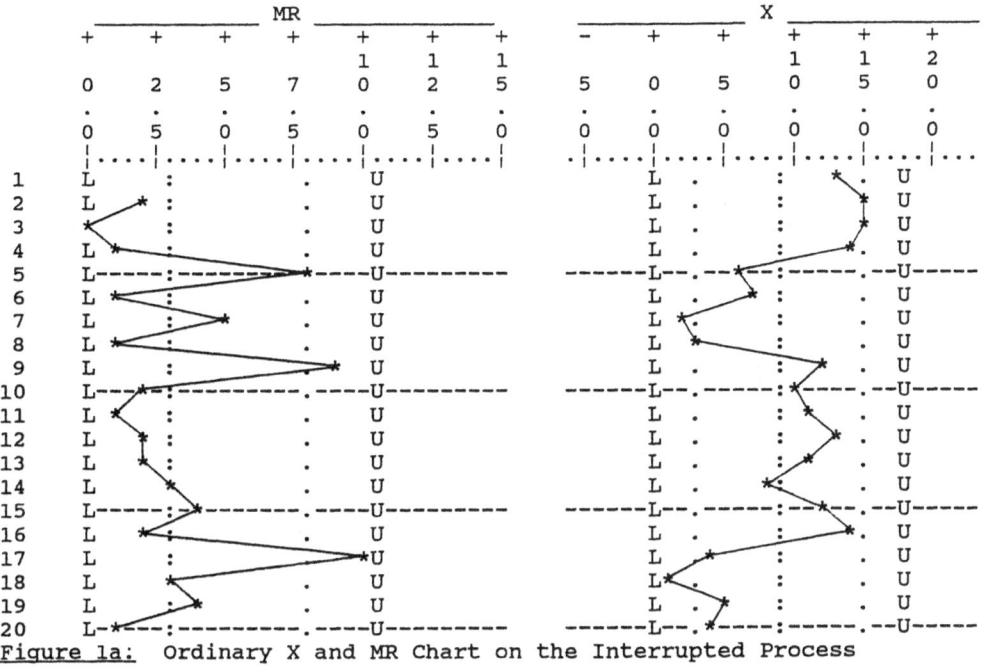

Figure 1a: Ordinary X and MR Chart on the Interrupted Process

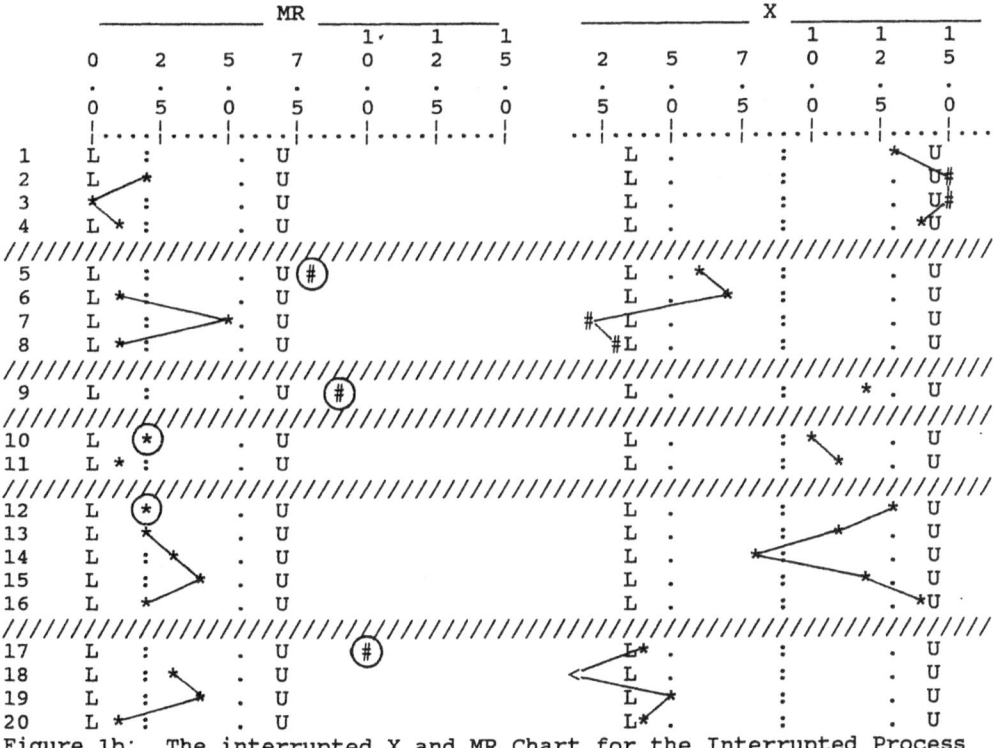

Figure 1b: The interrupted X and MR Chart for the Interrupted Process

The MR values following a process interruption have been circled. The centerline was calculated without using these MR values, but plotting them with a special symbol shows their systematically high nature compared to the within-run moving ranges. Valid points within the same run on both the MR and X charts have been connected, rendering the within-run uniformity clearly visible.

The requirement for a long uninterrupted production run for the ordinary X and MR chart has been circumvented by the simple expedient of dropping the initial moving range from each run. It may also be noted that the requirement for 100 percent inspection also vanishes here. Any time-ordered subset of the short runs may be selected for charting purposes and still not violate the mandate for homogeneity within each moving subgroup. In the next section a special case of the interrupted X and MR chart will be considered which eliminates the difficulties in interpretation caused by lack of independence between successive moving ranges.

6. THE "RX" CHART FOR TIME-ORDERED DATA

The interrupted X and MR chart of the previous section was applicable for any or all of a series of short runs, where the breaks between these runs were thrust upon the quality engineer by the process, rather than being introduced at his option. In this section the method will be extended to the special case where the sampling is deliberately done in the same prescribed manner that would be used for an $\bar{X}$ and R chart with subgroups of size two. The two consecutive pieces that make up the sampling subgroups are selected with the proviso that each subgroup be as homogeneous as possible which requires that both pieces come from the same production run.

The resulting data will have the following pattern:

X, X / X, X / X, X / X, X / X, X / X, X / X, X.

The pairs of consecutive observations may be selected in the manner best suited to the analysis of the process. Whether the process is a single long run or a series of short runs, 100 percent inspection is no longer needed to get homogeneous subgroups. Furthermore, by deleting the "moving" ranges for the first observation in each subgroup, only the independent ranges of subgroups of size two remain, the same ranges that would have existed for an $\bar{X}$ and R chart with subgroups of size two. The problem of interpreting the ordinary MR chart, due to lack of independence between the moving range, has been completely eliminated.

The Shewhart-type control chart described here has been named the RX chart [Hart and Hart 1989 pp. 255-258]. It may be looked upon as a special case of the interrupted X and MR chart, or as a modification of the $\bar{X}$ and R chart with subgroups of size two where the individuals instead of the means of the subgroups are plotted. The control limits may readily be shown to be:

3σ limits.

$$LCL(X) = \bar{X} - 2.66\,\bar{R} \qquad UCL(X) = \bar{X} + 2.66\,\bar{R}$$
$$LCL(R) \quad None \qquad UCL(R) = 3.27\,\bar{R}$$

2σ limits.

$$LCL(X) = \bar{X} - 1.77\,\bar{R} \qquad UCL(X) = \bar{X} + 1.77\,\bar{R}$$
$$LCL(R) \quad None \qquad UCL(R) = 2.51\,\bar{R}$$

The advantages of the RX chart over the usual $\bar{X}$ and R chart for time ordered data include the following:

1. The RX chart is simpler to prepare. No subgroup means have to be calculated and the ranges of two values are easier to calculate than are the ranges of four or five values.

2. Any peculiarities of individuals are not masked by the use of subgroup means.

3. The points plotted on the X chart may be compared directly with specification tolerance limits, avoiding the confusion which arises when people inevitably try to compare the $\bar{X}$ values with the specifications.

4. With the RX chart, typically fewer observations are required to detect a change in the process, as predicted by Shewhart.

This final point of the higher efficiency of the RX chart is substantiated in the next section. RX chart results from 15 data sets from the literature are compared to the original control charts which used averages.

7. EXAMPLES OF RX CHARTS

A total of 15 quality control data sets with time-ordered data and subgroup sizes of four or five were taken from the literature dating from 1931 to 1986. Each of the data sets had been analyzed with an $\bar{X}$ and R chart except for Shewhart's 1931 data for which he used an $\bar{X}$ and σ chart. Each example showed evidence that the process was not stable over time. When all of the original data was used, the RX chart always did as well or better at detecting a change in the process than did the original chart. In many cases only the first two observations from the original subgroups of four or five were used for the RX chart and it still compared favorably with the original analysis which used all of the data.

#	Author Year	(Ref. #) Pages	Original		RX	RX Results compared to original charts
			n	N	N	$(\bar{X}$ & R or $\bar{X}$ & $\sigma)$
1	ASTM 1976	(2) 129-131	4	32	32	more indications sooner
2	ANSI 1985	(1) 7	4	40	40	same as original
3	Ott 1975	(11) 51-53	5	125	50	earlier indications with 40% of data
4	Ott 1975	(11) 60-61	4	96	48	same indications with 50% of data
5	Shewhart 1931	(12) 20-21, 313	4	204	102	same indications with 50% of data
6	Bicking & Gryna 1974	(3) Chap 13 52-53	4	100	50	same indications with 50% of data
7	Burr 1976	(4) 68-70	4	100	50	same indications with 50% of data
8	Burr 1976	(4) 69-70	4	100	50	50% of indications with 50% of data
9	Burr 1979	(5) 162-163	5	100	40	same indications with 40% of data
10	Burr 1979	(5) 162-163	5	100	40	same indications with 40% of data
11	Grant & Leavenworth 1980	(8) 9-11	5	100	40	same indications with 40% of data
12	Grant & Leavenworth 1980	(8) 140-141	5	80	32	same indications with 50% of data
13	Montgomery 1985	(10) 177-179	5	125	50	same indications with 40% of data
14	Montgomery 1985	(10) 182-183	5	75	30	same indications with 40% of data
15	Wheeler & Chambers 1986	(15) 57	4	144	72	same indications with 50% of data

Table 1: Summary of Results for RX Charts Compared to $\bar{X}$ and R Charts.

A summary of the results from these 15 data sets are shown in Table 1. The original subgroup size, n, and the total number of original observations, N, as well as the total number of observations used for the RX chart are shown in the table.

The original control chart and the RX chart are shown below for three of the 15 data sets which were considered to be of particular interest.

>Figures 2a and 2b are from an ASTM [1976 pp. 129-131] example. This example was selected because the published $\bar{X}$ and R chart was supplemented by a chart for individuals which the publication noted to be very powerful, compared to the chart for averages.

>Figures 3a and 3b are from the ANSI [1985 p. 7] example, chosen because it shows as poor a performance for the RX chart as was found. Ten subgroups of size four were used by ANSI.

>Figures 4a and 4b are from Ott [1975 pp. 51-53] and shows the most favorable comparison of the chart for individuals to the chart for averages from the 13 data sets studied.

ASTM Example 23 with 8 subgroups of size four is shown in Figure 2a as an $\bar{X}$ and R chart. The auxiliary chart for the 32 consecutive individuals is shown in Figure 2b. Standard values of 20 for $\bar{X}$ and .9 for σ were given. Commenting on the charts the ASTM states:

>At the outset both the chart for ranges and the chart for individuals (but not the chart for averages) gave indications of lack of control. Subsequently for sample 6 (individuals 21 through 24) the control chart for individuals showed the first unit in the sample of 4 to be outside the upper control limit, thus indicating lack of control before the entire sample was obtained.

The superiority of the chart for individuals over the chart for averages on time-ordered data is consistent with what has been found elsewhere and should not be surprising.

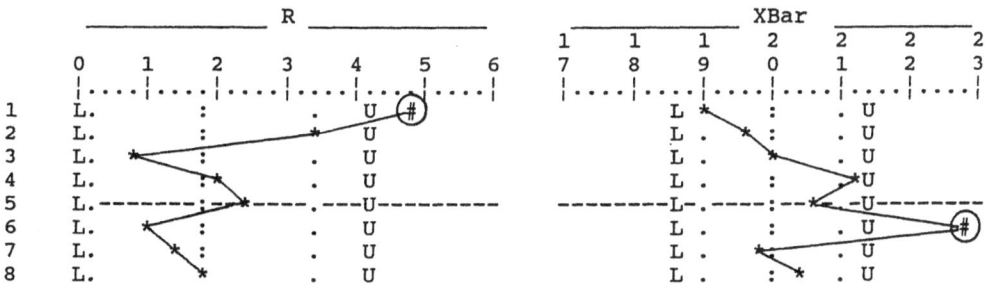

__Figure 2a:__ The Original $\bar{X}$ and R Chart for ASTM's Example 23

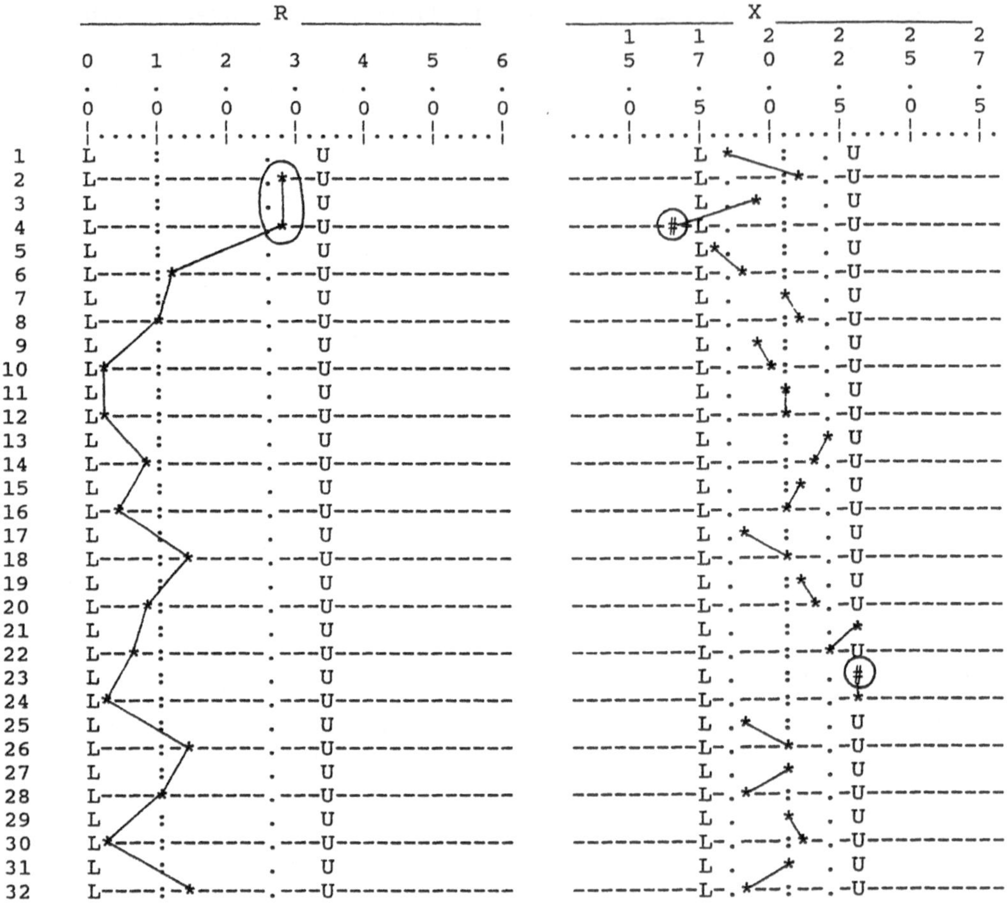

132

```
                   R                                        X
       0   1   2   3   4   5   6              1   1   2   2   2   2
                                              5   7   0   2   5   7
       0   0   0   0   0   0   0              0   5   0   5   0   5
       |...|...|...|...|...|...|              ...|...|...|...|...|...|.
 1     L   :       .   U                      L *   :   . U
 2     L---:-------.*--U-------               -------L-.---:-*-.--U--------
 3     L   :       . |U                       L   .-----* :   . U
 4     L---:-------.*--U-------               -(#)L-.---:---.--U--------
 5     L   :     * .   U                      L*.       :   . U
 6     L---:---*---.---U-------               -------L-.---:---.--U--------
 7     L   :       .   U                      L   .---* :   . U
 8     L---:-*-----.---U-------               -------L-.---:-*-.--U--------
 9     L   :       .   U                      L   . *   :   . U
10     L*--:-------.---U-------               -------L-.---*:---.--U--------
11     L   :       .   U                      L   .   * :   . U
12     L*--:-------.---U-------               -------L-.---:-*-.--U--------
13     L   :       .   U                      L   .   :   *.  U
14     L---*:------.---U-------               -------L-.---:-*-.--U--------
15     L   :       .   U                      L   .   :*  .   U
16     L-*-:-------.---U-------               -------L-.---*---.--U--------
17     L   .       .   U                      L   .*--:   .   U
18     L---:--*----.---U-------               -------L-.---*---.--U--------
19     L   :       .   U                      L   .   :*  .   U
20     L---:-*-----.---U-------               -------L-.---:-*-.--U--------
21     L   :       .   U                      L   .   :   .*  U
22     L--*:-------.---U-------               -------L-.---:---*U--------
23     L   :       .   U                      L   .   :   .(#)
24     L*--:-------.---U-------               -------L-.---:---.*--------
25     L   :       .   U                      L   .*  :   . U
26     L--:--*-----.---U-------               -------L-.---*---.--U--------
27     L   :       .   U                      L   .   :*  . U
28     L---:--*----.---U-------               -------L-.*--:---.--U--------
29     L   :       .   U                      L   .   *   . U
30     L*--:-------.---U-------               -------L-.---*-.---U--------
31     L   :       .   U                      L   .   /* . U
32     L---:-*-----.---U-------               -------L-.*--:---.--U--------
```

Figure 2b: RX Chart for ASTM's Example 2

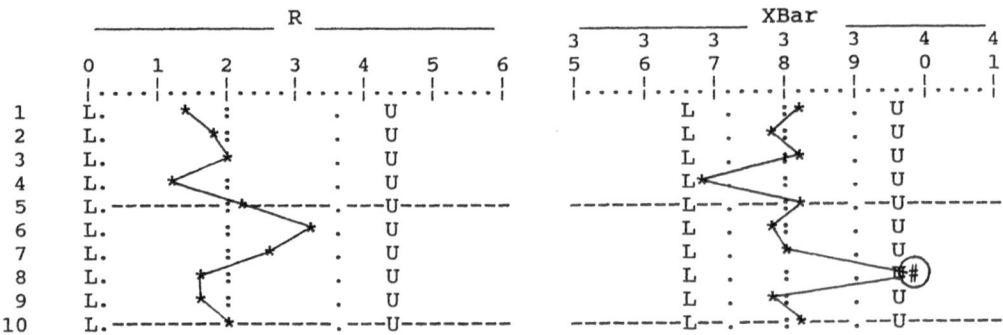

```
                   R                                     XBar
       0   1   2   3   4   5   6              3   3   3   3   3   4   4
                                              5   6   7   8   9   0   1
       |...|...|...|...|...|...|              |...|...|...|...|...|...|
 1     L. *   :       .   U                   L   .   :   .   U
 2     L.   *:        .   U                   L   .   *   .   U
 3     L.   :*        .   U                   L   .   :*  .   U
 4     L.   :         .   U                   L*  :   .   U
 5     L.---:---*-----.---U-------            ---L--.---:-*-.---U-------
 6     L.   :       * .   U                   L   .--*:   .   U
 7     L.   :     * .   U                     L   .   *   . U
 8     L.  *:         .   U                   L   .   :   .*(#)
 9     L.  *:         .   U                   L   .   *   . U
10     L.-------*-----.---U-------            ---L--.---:-*---.--U-------
```

Figure 3a: Original $\bar{X}$ and R Chart for ANSI Example

The RX chart for the same 32 observations is shown in Figure 2b. The evidence of lack of statistical uniformity within the dispersion of the first four individuals was detected by a single point out of the 3σ limits on the original R chart in Figure 2a and by two consecutive R values outside of the same 2σ limit in the R chart in Figure 2b. The RX chart in Figure 2b detects the same lack of statistical uniformity of the individuals noted by the ASTM comments as well as a run of 7 (out of only 32) individuals above the centerline.

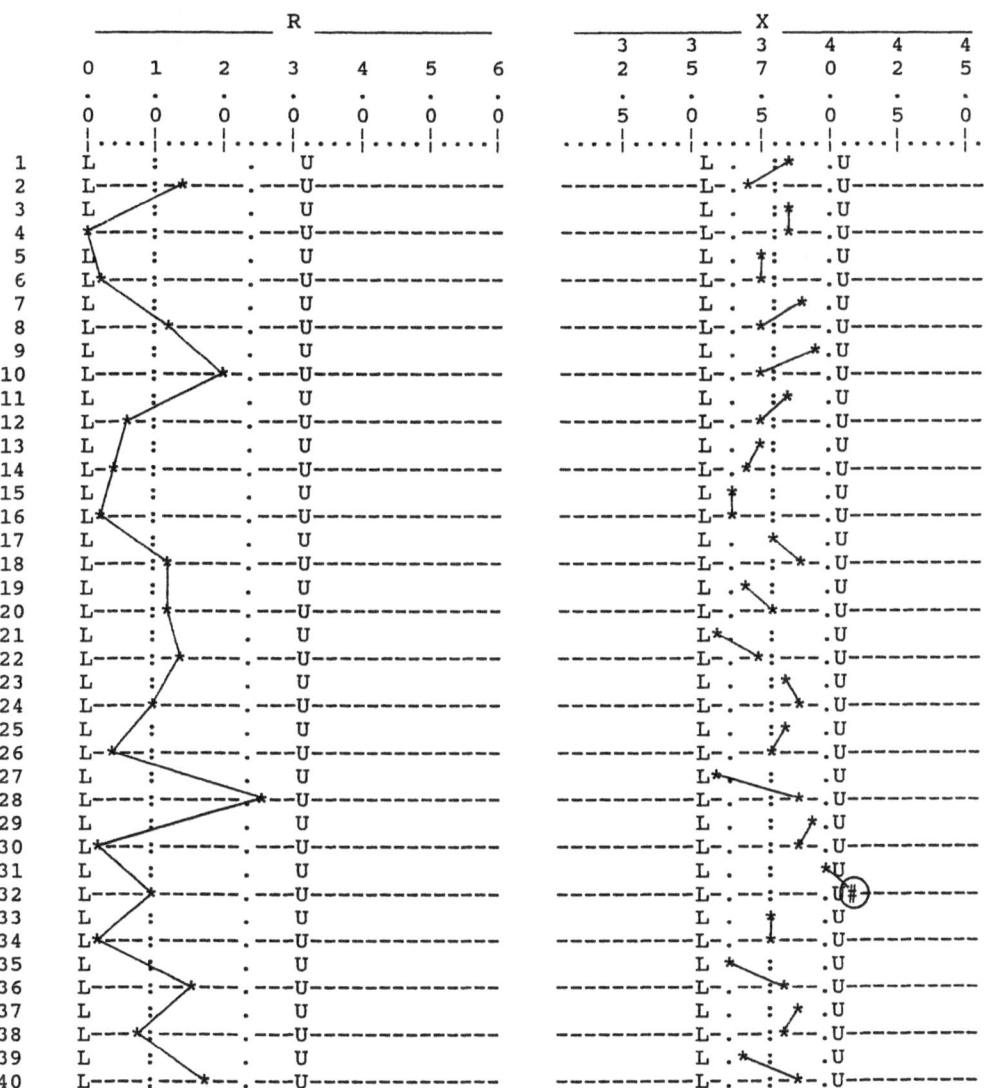

Figure 3b: RX chart for the ANSI Example

The second example shows the RX chart in as poor a light as any of
the 15 data sets examined. The original ANSI $\bar{X}$ and R chart for 10
subgroups of size four is shown in Figure 3a. The corresponding RX
chart for the 40 observations is shown in Figure 3b. A comparison of
the two figures shows no difference between the efficiency of the two
charts in detecting the process changes that occurred over time. The
third and final example shows the RX chart at its best when compared to
the $\bar{X}$ and R chart. Figure 4a shows the $\bar{X}$ and R chart for Ott's 25
subgroups of size 5. Lack of stability was evidenced by long runs of
eight $\bar{X}$ values which occurred below the centerline at subgroup 16 and
above the centerline at subgroup 24.

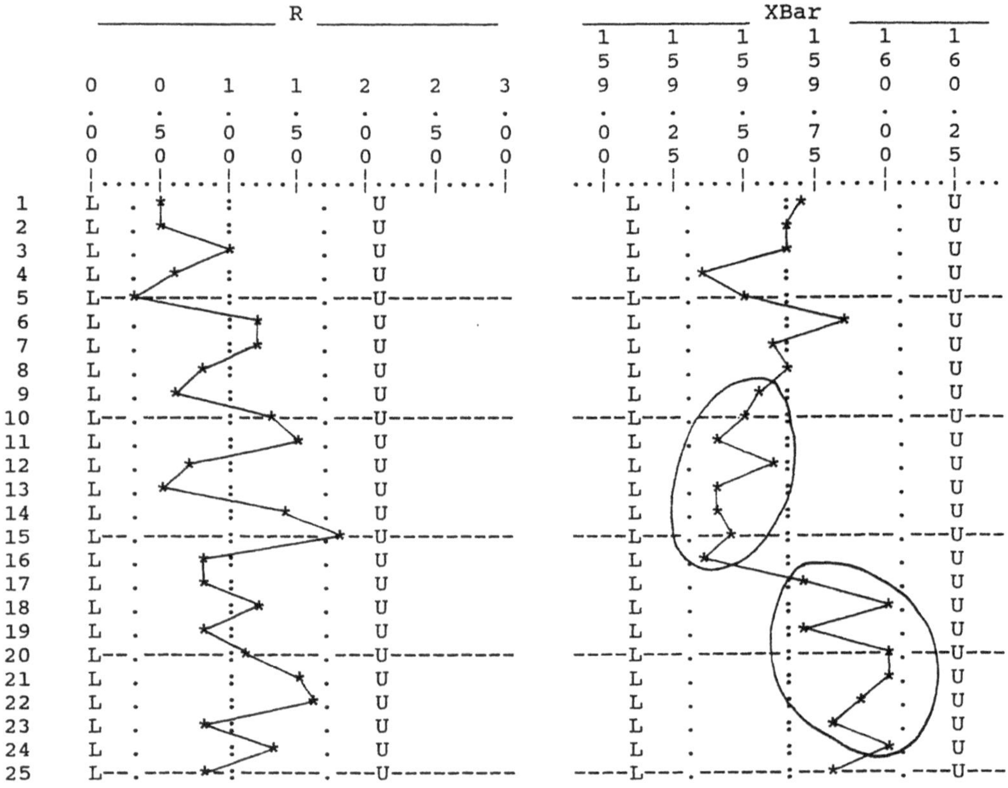

```
                 _____ R _____              _____ XBar _____
                                                            1     1     1     1     1     1
                                                            5     5     5     5     6     6
           0     0     1     1     2     2     3             9     9     9     9     0     0
           .     .     .     .     .     .     .            .     .     .     .     .     .
           0     5     0     5     0     5     0            0     2     5     7     0     2
           0     0     0     0     0     0     0            0     5     0     5     0     5
           |...|...|...|...|...|...|                      ..|...|...|...|...|...|...
  1      L . *    :       .  U                             L  .       :*      .  U
  2      L . *    :       .  U                             L  .         *      .  U
  3      L .        *     .  U                             L  .       *       .  U
  4      L . *     :       .  U                            L  . *     :       .  U
  5      L--*-----:-------.---U--------                   ---L---.---*--:-------.---U---
  6      L . :    *  .  U                                  L  .     :       *   .  U
  7      L . :   *   .  U                                  L  .     *:      .  U
  8      L . *  :    .  U                                  L  .       *:    .  U
  9      L . *    :   .  U                                 L  .     *:      .  U
 10      L--.-----:--*--.---U--------                     ---L---.--*--:----.---U---
 11      L . :    *   .  U                                 L  . *    :      .  U
 12      L . :     *  .  U                                 L  .   *  :      .  U
 13      L . *  :     .  U                                 L  .  *   :      .  U
 14      L . :     *  .  U                                 L  .  *:        .  U
 15      L--.-----:----*--U--------                       ---L--.---*--:-------.---U---
 16      L . *  :      .  U                                L  . *    :      .  U
 17      L . :  *      .  U                                L  .      :       *  .  U
 18      L . :  *      .  U                                L  .      :     *   .  U
 19      L . *  :      .  U                                L  .      :  *      .  U
 20      L--.-----:--*--.---U--------                     ---L---.-----:----*--.--U---
 21      L . :     *  .  U                                L  .      :       *  .  U
 22      L . :    *   .  U                                L  .      :    *     .  U
 23      L . *  :      .  U                                L  .      :     *    .  U
 24      L . :   *   .  U                                 L  .      :     *    .  U
 25      L--.-----*--:-------.---U--------                ---L---.------:--*----.---U---
```

Figure 4a: The original $\bar{X}$ and R Chart for Ott's Example

Figure 4b shows the RX chart which results when only the first two
observations are retained from each subgroup. X value number 22, the
second observation in Ott's subgroup 11, was below the lower 3σ limit.
X value number 44, the second value in Ott's subgroup 22, was above the
upper 3σ limits. Although only 40 percent of the original data was

135

used, the chart for individuals gave earlier indications of the process
instability than did the original chart for averages.

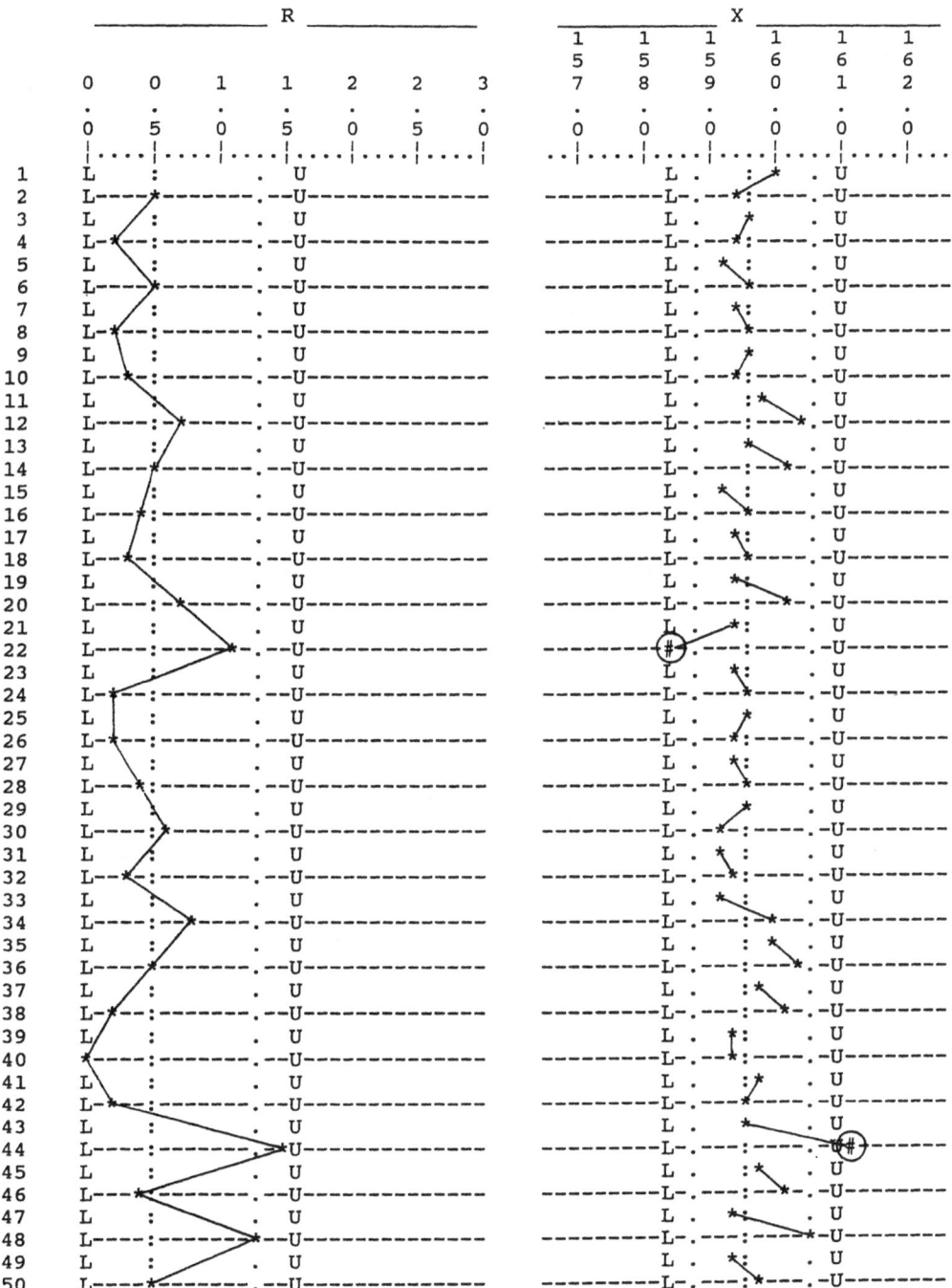

Figure 4b: The RX Chart for Ott's Example

8. SUMMARY

As originally noted by Shewhart [1931], control chart analysis of time-ordered data would best be accomplished by charting the individual values. This has since been accomplished with the X and MR chart, but not without significant limitations and difficulties. The interrupted X and MR chart described in this paper overcomes some of these difficulties in addition to providing a simple method of coping with short production runs. A special case of the interrupted X and MR chart, the RX chart, is applicable to a wide range of processes. It overcomes the restrictions of the X and MR chart while meeting Shewhart's original suggestion of a control chart for individuals.

9. REFERENCES

[1] AMERICAN NATIONAL STANDARD (1985): ANSI Standards Z1.1 Guide for
     Quality Control; Z1.2 Control Chart Method for Analyzing
     Data; and Z1.3 Control Chart Method of Controlling Quality
     During Production.
     New York: American National Standards Institute.

[2] AMERICAN SOCIETY FOR TESTING AND MATERIALS (1976): ASTM Manual on
     Presentation of Data and Control Chart Analysis.
     Philadelphia: American Society for Testing and Materials.

[3] BICKING, C. A. and GRYNA, F. M. (1974): Process Control by
     Statistical Methods, in Quality Control Handbook 3rd edition
     Editor J. M. Juran. New York: McGraw-Hill.

[4] BURR, I. W. (1976): Statistical Quality Control Methods.
     New York: Marcel Dekker.

[5] BURR, I. W. (1979): Elementary Statistical Quality Control.
     New York: Marcel Dekker.

[6] DEMING, W. E. (1982): Quality, Productivity, and Competitive
     Position.
     Cambridge: Massachusetts Institute of Technology, Center for
     Advanced Engineering Studies.

[7] DEMING, W. E. (1986): Out of the Crisis.
     Cambridge: Massachusetts Institute of Technology, Center for
     Advanced Engineering Studies.

[8] GRANT, E. L. and LEAVENWORTH, R. S. (1980): Statistical Quality
     Control.
     New York: McGraw-Hill.

[9] HART, M. K. and HART, R. F. (1989): Quantitative Methods for
     Quality and Productivity Improvement.
     Milwaukee: American Society for Quality Control Quality
     Press.

[10] MONTGOMERY, D. C. (1985): Introduction to Statistical Quality
     Control.
     New York: John Wiley and Sons.

[11] OTT, E. R. (1975): Process Quality Control.
     New York: McGraw-Hill.

[12] SHEWHART, W. A. (1931): Economic Control of Quality of
     Manufactured Product.
     New York: D. Van Nostrand.

[13] SHEWHART, W. A. (1943): Statistical Control in Applied Science,
     Transactions of the American Society of Mechanical
     Engineering, Cleveland, April, 222-225.

[14] WESTERN ELECTRIC COMPANY (1956): Statistical Quality Control
     Handbook, Editor B. Small.
     Newark: Western Electric Company.

[15] WHEELER, D. J. and CHAMBERS, D. S. (1986): Understanding
     Statistical Process Control.
     Knoxville: Statistical Process Controls.

On an Adaptive Acceptance Control Chart
for Autocorrelated Processes

P. Thyregod and H. Madsen, Lyngby, Denmark

ABSTRACT

Standard control chart techniques for process surveillance are usually assessed under the assumption that sampled values from a process in control may be represented by independent, identically distributed random variables. In some applications, however, the process exhibits an inherent dynamics with slow variations of the process level, even in a state of statistical control. In order to describe these variations we suggest a simple time–series model for the process level. The model implies that an explicit estimate of the current process level may be obtained by the Kalman Filter technique. In the paper we show how this estimate may be utilized in an acceptance control chart to vary the sample size in accordance with the estimated actual process level, without sacrificing a specified consumer protection.

1. INTRODUCTION

Control charts were originally developed as a tool to verify whether a process is in a state of statistical control, i.e., whether the process appears to be subject only to normal disturbances, or whether an "assignable cause" has occurred, calling for further investigation with the purpose of removing this cause for good.

In classical control charting approaches, this state of statistical control is generally considered to be equivalent with the assumption that samples from succesive subgroups behave like IID random variables. Even for charts that intend to estimate some "current process level" like EWMA-charts, or GMA–charts, succesive samples are often considered to behave like independent variables with a fixed, deterministic structure of the process mean.

Some later developments of control chart techniques, like e.g. the acceptance control chart, have widened the scope from verifying this simple state of statistical control to verifying whether the process output may be considered to be within specified limits, or not. Such approaches allow for some latitude in the interpretation of the concept of statistical control in the sense that process level is allowed to fluctuate, as long as process level stays within a certain band of acceptable process levels. When assessing the performance of acceptance control charts it is, however,

Frontiers in Statistical Quality Control 4
Ed. by Lenz et al.
© Physica-Verlag Heidelberg 1992

common practice to assume that samples from succesive subgroups are independently distributed.

In some applications, this assumption of independence between successive sample results is only a very crude approximation. Thus, for many chemical and environmental processes the "natural variation" is not pure noise, but the variation includes an inherent dynamical component. The process dynamics is observed as a strong autocorrelation between successive sample results. In such situations a stationary time series model that includes the process dynamics will often give a better description of the process performance in the in−control state than the simpler model with independent observations.

Once the time series model has been identified, the current state of the process may be estimated by means of an adequate filtering technique, and moreover the model may be used for predicting future states of the process.

In the paper we shall consider a situation that would traditionally call for application of an acceptance control chart with a specified consumer's risk: The parameter of interest is the fraction of production that exceeds some specification limit. It is desired that the inspection procedure will detect a given proportion, p_1 of nonconforming units with a specified probability $1-\beta$.

Traditionally, such a chart is designed with a fixed subgroup sample size that has been determined in order to yield a sufficiently large acceptance probability for a specified value $p_0 < p_1$, of the proportion of nonconforming product. It is, however, generally of interest to avoid excessive sampling, and therefore it is desirable to use a varying sample sample size that is adjusted in accordance with the most recent observations. In the paper we indicate how the estimate of the current state of the process, and the prediction of future process values may be utilized to minimize the size of the next sample without violating the requirements on the detection probability.

Reynolds et al. (1988), (1989) have suggested a procedure for the design of a Shewhart $\bar{X}$−chart with a sampling interval that varies in accordance with the result of the most recent sample. The fundamental difference between the two approaches is that the procedure suggested by Reynolds et al. is designed under consideration of the average run length corresponding to a fixed process level $\mu(t) \equiv \mu$ whereas our approach uses an explicit estimation of the current process level $\mu(t)$ to assess the size of the next subgroup sample.

2. THE PROCESS MODEL

We shall consider a situation where a process is sampled at regular time intervals, and measurements, $X_1(t), X_2(t), \ldots, X_{n_t}(t)$ of the process output, e.g. filling weight, representative for the output at time t, are recorded.

We shall assume that the variation of the measurements $X_i(t)$ within a sample (subgroup) may be described by *the observation equation*

$$X_i(t) = \mu(t) + \epsilon_i(t) \tag{1}$$

where $\epsilon_i(t)$, $i=1,2...,n_t$; $t=1,2,....$ are mutually independent $N(0,\sigma^2)$ –distributed random variables.

In the following the subgroup mean, $\mu(t)$, will be termed *the process level*, and σ^2 the *within subgroup variance*.

In most applications of control chart techniques, a process is considered to be in control, if the process level $\mu(t)$ does not vary between subgroups, i.e. if the sample averages

$$\bar{X}(t) = \{X_1(t)+X_2(t)+\cdots+X_{n_t}(t)\} \ / \ n_t \ , \ t = 1,2,..$$

may be described by mutually independent Normally distributed variables with a common mean and with variance σ^2/ n_t .

In many applications it seems, however, appropriate to extend this concept of statistical control to include stationary processes. Thus, it might be more appropriate to consider a process to be in statistical control, if the variation of process level $\mu(t)$ over time may be described by a *linear stochastic process*. In the present exposition we shall assume the most simple model for $\mu(t)$, viz.

$$\mu(t) = \alpha \, \mu(t-1) + u(t) \tag{2}$$

where $u(t)$, $t= 1,2,...$ are mutually independent $N(0, \sigma_0^2)$ – distributed random variables, and where the parameter α denotes the autocorrelation between successive values of the process level $\mu(t)$. The model (2) is a so called *autoregressive model of order 1* (AR(1)).

We note that when all subgroup samples are of the same size, $n_t = n$, then it follows from the assumptions (1) and (2) that the sequence of sample averages

$$\bar{X}(t) = \{X_1(t)+X_2(t)+\cdots+X_n(t)\} \ / \ n \ , \ t = 1,2,..$$

may be described by an autoregressive moving average (ARMA) process of order (1,1), see Harvey (1981).

The model is a special case of a more general class of models. The important point is that we have separated the model for the variation in process output into two components, an observation equation (1), and a system equation (2) expressing the variation of process level $\mu(t)$ in terms of a first order Markov process. In general we may have

$$\mu(t) = \mathbf{H} \, \mathbf{z}(t) \tag{3}$$

where $\mathbf{H}$ is a $1 \times k$ vector of fixed coefficients, and the $k \times 1$ vector $\mathbf{z}(t)$ satisfies a linear stochastic difference equation, *the system equation:*

$$z(t) = F\, z(t-1) + w(t) \tag{4}$$

where F is a $k \times k$ matrix of fixed coefficients and $w(t)$ is a sequence of independent, normally distributed random $k \times 1$ vectors with mean zero and variance–covariance matrix W

The process $z(t)$ described by (4) is a first order Markov vector process. Hence, all information about future values $z(t+\tau)$ with $\tau > 0$ are contained in the present value $z(t)$. Therefore the vector $z(t)$ is often termed a *state vector*, and the representation by (1), (3) and (4) is termed a *state space representation*.

3. THE KALMAN FILTER FOR ESTIMATION OF CURRENT PROCESS LEVEL $\mu(t)$

We shall return to the model given by (1) and (2), and describe the estimation of the process level for subgroup t following observations $x_1(1), x_2(1), \ldots, x_{n_1}(1)$; $x_1(2), x_2(2), \ldots, x_{n_2}(2)$; $\ldots$; $x_1(t), x_2(t), \ldots, x_{n_t}(t)$ with the corresponding sample averages $\bar{x}(1), \bar{x}(2), \ldots, \bar{x}(t)$

Since subgroups are sampled successively, we shall consider the estimation of $\mu(t)$ by the recursive updating of the estimates $\mu(1), \mu(2), \ldots, \mu(t-1)$ in light of the succesive observations $\bar{x}(1), \bar{x}(2), \ldots, \bar{x}(t)$ of the sample average.

For a given subgroup, t, observation of the current subgroup average, $\bar{X}(t) = \bar{x}(t)$, gives rise to an estimate $\hat{\mu}_f(t)$ of the current value of the process level, $\mu(t)$. The estimate will in general be based upon the current process history, i.e. observation of all the subgroup averages up to, and including $\bar{X}(t)$. The estimate $\hat{\mu}_f(t)$ is termed the *filtered* value .

The filtered value may be used as starting point for prediction of future values of the process level. Here we shall only be concerned with predicting the value $\mu(t+1)$ for the next period $t+1$. We shall use the symbol $\hat{\mu}_p(t+1)$ to denote this predicted value.

The succesive updating of the estimates $\hat{\mu}_f(t)$ and $\hat{\mu}_p(t+1)$ for $t = 1, 2,$ is given by the Kalman-filter technique. An introduction to this technique may be found, for example, in the textbook by Jaswinski (1970). Given the quantities $\hat{\mu}_f(t-1)$ and $\gamma(t-1)$, and $w(t-1)$ an iteration of the Kalman Filter entails the following equations

$$\hat{\mu}_p(t) = \alpha\, \hat{\mu}_f(t-1) \tag{5}$$

$$\gamma(t) = \alpha^2 \frac{1}{n_{t-1} + 1/\gamma(t-1)} + \gamma_0 \tag{6}$$

$$w(t) = \frac{n_t}{n_t + 1/\gamma(t)} \tag{7}$$

$$\hat{\mu}_f(t) = \{1-w(t)\} \, \hat{\mu}_p(t) + w(t) \; \bar{x}(t) \tag{8}$$

with

$$\gamma_0 = \sigma_0^2 / \sigma^2$$

Usually the filter is seeded with

$$\hat{\mu}_f(0) = 0 \qquad \text{and} \qquad \gamma(0) = 0$$

such that the first iteration yields

$$\hat{\mu}_p(1) = 0 \qquad \text{and} \qquad \gamma(1) = a^2 \frac{n_1}{n_1 + 1/\gamma_0} + \gamma_0$$

In Kalman filter theory, the weight $w(t)$ is usually denoted the *Kalman Gain*. The precision parameter $\gamma(t)$ expresses the accuracy of the state prediction in (5) in terms of reciprocal "equivalent sample size".

If the subgroup sample size, n_t , is kept constant, it follows that the precision parameters $\gamma(t)$ will tend to a limiting value determined as the stationary solution, $\gamma(\infty)$ to (6). Thus, the Kalman Gain, $w(t)$, will tend to a limiting value $w(\infty)$ given by

$$w(\infty) = \frac{n}{n + 1/\gamma(\infty)}$$

Rewriting the filtering equation (8) as

$$\hat{\mu}_f(t) = \hat{\mu}_p(t) + w(t)\{ \bar{x}(t) - \hat{\mu}_p(t)\} \tag{9}$$

we therefore find that, in the limit, the successive filtering is equivalent with an exponentially weighted moving average, or geometric moving average, of subgroup means as noted, e.g. by Hunter (1986).

The successive filtering and prediction may be interpreted by means of conditional distributions in analogy with the Bayesian updating mechanism. Following Meinhold and Singpurwalla (1983) we obtain:

a)   Before observation of $\bar{X}(t)$ the prior knowledge about $\mu(t)$ given $\bar{x}(1)$, $\bar{x}(2)$, ... , $\bar{x}(t-1)$ may be represented by a normal distribution with mean $\hat{\mu}_p(t)$ and variance $\sigma_p^2(t)$ with

$$\sigma_p^2(t) = \gamma(t) \, \sigma^2 \tag{10}$$

b) It follows from the observation equation (1) that the distribution of $\bar{X}(t)$ for a given $\mu(t)$ (and $\bar{x}(1)$, $\bar{x}(2)$, ... ,$\bar{x}(t-1)$) is a normal distribution with mean $\mu(t)$ and variance σ^2/n_t.

c) After observation of $\bar{X}(t) = \bar{x}(t)$, the posterior distribution of $\mu(t)$, (i.e. the conditional distribution of $\mu(t)$ given $\bar{X}(t) = \bar{x}(t)$, and given $\bar{x}(1)$, $\bar{x}(2)$, ... ,$\bar{x}(t-1)$) is a normal distribution with mean $\hat{\mu}_f(t)$ and variance $\sigma_f^2(t)$, where

$$\sigma_f^2(t) = \sigma^2 \frac{1}{n_t + 1/\gamma(t)} \simeq \frac{\sigma^2}{n_t}\, w(t) \qquad (11)$$

d) After observation of $\bar{X}(t) = \bar{x}(t)$, the conditional distribution of $\mu(t+1)$ given $\bar{X}(t) = \bar{x}(t)$, and given $\bar{x}(1)$, $\bar{x}(2)$, ... ,$\bar{x}(t-1)$ is a normal distribution with mean $\hat{\mu}_p(t) = \alpha\, \hat{\mu}_f(t)$ and variance

$$\sigma_p^2(t+1) = \alpha^2 \sigma_f^2(t) + \sigma_0^2$$

$$= \sigma^2 \{ \frac{\alpha^2}{n_t}\, w(t) + \gamma_0 \} \qquad (12)$$

This distribution describes the information about $\mu(t+1)$ that is available prior to observation of $\bar{X}(t+1)$. Hence, we may proceed with a) .

4. THE ACCEPTANCE CONTROL CHART WITH FIXED CONSUMER'S RISK

We shall consider a situation where it would be appropriate to use an acceptance control chart with a specified consumer protection, as the chart suggested by Freund (1957). Such a situation might occur in the control of various filling processes, in the control of mandatory standards controlled by external agencies, etc.

We shall assume that a lower specification limit, L, has been given for the product quality characteristic, X, and that the variable of interest is the proportion, p, of product from a subgroup that does not conform to this specification. We shall furthermore assume that a *rejectable proportion*, p_1, of nonconforming product has been specified.

The process is sampled at regular time intervals in order to verify that the current process output is satisfactory, i.e., that the current proportion of nonconforming product does not exceed p_1. If the proportion of nonconforming product is too large, the current output is not accepted, and various corrective actions may be initiated.

The consumer's risk for the acceptance control chart is defined in analogy with lot by lot sampling procedures as the probability of accepting current process output when the true proportion of nonconforming product is p_1. The smaller the value of β, the better is the protection against unwanted acceptance of unsatisfactory product. We shall here consider procedures with a specified consumer's risk, β.

From the well–known relation between current process mean, μ, (in this section only a single time instant is considered, and therefore the index t has been suppressed) and fraction nonconforming, p, in the subgroup,

$$\mu_p = L + \sigma \, z_{1-p}$$

we obtain the relation between .the rejectable proportion, p_1 , and the corresponding *rejectable process level (RPL)* as

$$RPL = \mu_{p_1} = L + \sigma \, z_{1-p_1} \tag{13}$$

where, as usual, z_p denotes the $100 \times p$ percentile in the standardized normal distribution, and σ^2 denotes the within subgroup variance.

Given the subgroup sample size, n, and the control limit, ACL, for the sample average, the acceptance criterion is "Accept if $\bar{X} \geq ACL$ ". When the current process level is μ, the corresponding probability of acceptance is determined as

$$P[\, \bar{X} \geq ACL \,|\, \mu \,] = 1 - \Phi \left[\frac{ACL - \mu}{\sigma/\sqrt{n}} \right] \tag{14}$$

Thus, specifying the subgroup sample size, n, and the risk, β , of accepting unsatisfactory product, the control limit shall satisfy

$$P[\, \bar{X} \geq ACL \,|\, \mu = RPL \,] = \beta \tag{15}$$

and hence the control limit is found as

$$ACL_n = RPL + z_{1-\beta} \, \sigma/\sqrt{n} \tag{16}$$

where the suffix n now is introduced to denote the dependence on the sample size.

When the acceptance procedure satisfies (15), it follows from (16) that the probability of acceptance corresponding to a given process level, μ, is given by

$$P[\, \bar{X} \geq ACL_n \,|\, \mu \,] = 1 - \Phi \left[\frac{ACL_n - \mu}{\sigma/\sqrt{n}} \right] = 1 - \Phi \left[z_{1-\beta} - \sqrt{n} \, \{\mu - RPL\}/\sigma \right] \tag{17}$$

The expression (17) demonstrates the well–known property of systems of acceptance sampling plans with a specified protection against unsatisfactory quality : If the acceptance criterion for a given subgroup sample size satisfies the restriction (15), then the probability of acceptance corresponding to a given process level. μ , with $\mu > \mu_{p_1}$ is an increasing function of the sample size, n.

In the design of acceptance control charts, it is usual practice to determine the sample size by considering some "acceptable proportion" p_0 of nonconforming product that usually should be accepted by the control procedure . Specifying p_0 we obtain the corresponding *acceptable process level (APL)* as

$$APL = \mu_{p_0} = L + \sigma \, z_{1-p_0}$$

If further one specifies the risk, α , of non–acceptance when $\mu = APL$, i.e.

$$P[\ \bar{X} \geq ACL_n \,|\, \mu = APL\ ] = 1 - \alpha \tag{18}$$

then the sample size may be determined from

$$n = \left\{ \frac{z_{1-\alpha} + z_{1-\beta}}{z_{1-p_0} - z_{1-p_1}} \right\}^2 \tag{19}$$

It is intuitively felt that if current process level is far above APL, then the sample size given by (19) will be larger than necessary in order to satisfy the specified risks. Therefore, if we have some indication that current process level is far above APL, we might *reduce* the amount of inspection, i.e., use a smaller subgroup sample size than (19), and, conversely, if current process level is below APL, while still satisfactory (i.e. above RPL), then we might use a larger sample size than (19) in order not to reject a satisfactory product.

In the following sections we shall discuss the determination of the sample size under consideration of the information about the current process level that has been determined by the Kalman filter.

5. DETERMINATION OF SAMPLE SIZE UNDER A CLASSICAL ACCEPTANCE CRITERION

Now consider the situation following inspection of subgroup t–1. Under the process model given by (1) and (2), we have that the conditional distribution of $\mu(t)$ given the sample history up to, and including subgroup t–1 is a normal distribution with mean $\hat{\mu}_p(t)$, and variance $\sigma_p^2(t)$ given by (10).

Therefore, when deciding upon the size of the sample from subgroup t it is natural to take this prior information into account, and balance the size of the sample with the available information on $\mu(t)$ without sacrificing the protection, $P[\text{ Accept}|\ \mu = RPL\ ]=\beta$, against acceptance of unsatisfactory quality that has been specified by (15) above.

The prior information will be used to assess the sample size that will be necessary to provide a sufficiently large probability of acceptance for satisfactory product. Thus, instead of (18) above, we shall consider

$$P[\ \bar{X}(t) \geq ACL_n|\ \mu(t) \geq RPL\ ] \geq 1-\alpha \tag{20}$$

where the conditioning event is evaluated in the prior distribution of $\mu(t)$ that is available before sampling subgroup t, i.e., in the conditional distribution of $\mu(t)$ given the process history $\bar{x}(1)$, $\bar{x}(2)$, ... , $\bar{x}(t-1)$.

In order to assess the conditional acceptance probability in (20) we shall need the following Lemma.

LEMMA 1

Consider the situation following inspection of subgroup $t-1$.
Let, for each value of n , the acceptance criterion satisfy

$$P[\ \bar{X} \geq ACL_n\ |\ \mu = RPL\ ] = \beta$$

Then

$$P[\ \bar{X}(t) \geq ACL_n\ ] \leq \beta\ \{\ 1 -P[\mu(t) > RPL]\ \} + P[\ \bar{X}(t) \geq ACL_n\ |\mu(t) > RPL]\ P[\mu(t) > RPL]$$

Proof:

Let $f(\mu)$ denote the density of $\mu(t)$ for given process history $\bar{x}(1)$, $\bar{x}(2)$, ... , $\bar{x}(t-1)$, prior to the observation of $\bar{X} \equiv \bar{X}(t)$, and consider the probability of accepting subgroup t

$$P[\ \bar{X} \geq ACL_n\ ] = \int P[\ \bar{X} \geq ACL_n\ |\mu]\ f(\mu)\ d\mu$$
$$= \int_{\mu \leq\ RPL} P[\ \bar{X} \geq ACL_n\ |\mu]\ f(\mu)\ d\mu + \int_{\mu > RPL} P[\ \bar{X} \geq ACL_n\ |\mu]\ f(\mu)\ d\mu$$

When the acceptance criterion ACL_n satisfies (15), it follows that $P[\ \bar{X} \geq ACL_n\ |\mu] \leq \beta$ for $\mu \leq RPL$, and therefore we have

$$\dot{P}[\ \bar{X} \geq ACL_n\ ] \leq \beta \int_{\mu \leq\ RPL} f(\mu)\ d\mu + \int_{\mu > RPL} P[\ \bar{X} \geq ACL_n\ |\mu]\ f(\mu)\ d\mu$$

The result in the Lemma then follows by noting that

$$\int_{\mu > RPL} P[\,\bar{X} \geq ACL_n \mid \mu\,]\; f(\mu)\; d\mu \;=\; P[\,\bar{X} \geq ACL_n \mid \mu > RPL\,]\; P[\mu > RPL]$$

$\square$

THEOREM 1

Consider the situation following inspection of subgroup t–1.

Let $\psi_1(n)$ be given by

$$\psi_1(n) = \kappa \sqrt{r\,n} - z_{\pi_1} \sqrt{n+r} - z_{1-\beta} \sqrt{r} \tag{21}$$

with

$$r = 1/\gamma(t)$$

$$\kappa = \{\,\hat{\mu}_p(t) - RPL\}/\sigma$$

and

$$\pi_1 = \beta\,\{1 - \Phi\,(\kappa \sqrt{r}\,)\,\} + (1-\alpha)\,\Phi\,(\kappa \sqrt{r}\,)$$

where $\hat{\mu}_p(t)$ denotes the prediction of $\mu(t)$ and $\gamma(t)$ the corresponding precision parameter determined by successive applications of (5) to (8).

If n_t satisfies

$$\psi_1(n_t) \geq 0$$

then

$$P[\,\bar{X}(t) \geq ACL_n \mid \mu(t) = RPL\,] = \beta$$

and

$$P[\,\bar{X}(t) \geq ACL_n \mid \mu(t) \geq RPL\,] \geq 1-\alpha$$

will hold for $n = n_t$.

Proof:

We first note that the conditional distribution of $\mu(t)$ given $\bar{x}(1),..,\bar{x}(t-1)$ is a normal distribution with mean $\hat{\mu}_p(t)$, and variance σ^2/r with $r = 1/\gamma(t)$. See a) in Section 3.
Therefore

$$P[\mu(t) > RPL] = \Phi\left[\,\frac{\hat{\mu}_p(t) - RPL}{\sigma \sqrt{1/r}}\,\right]$$

$$= \Phi\,(\kappa \sqrt{r}\,)$$

Thus, it follows from the Lemma 1 that if

148

$$P[\,\bar{X}(t) \geq ACL_n\,] \geq \pi_1 \tag{22}$$

then (20) will be satisfied.

It follows from a) and b) in Section 3 that the predictive distribution of $\bar{X}(t)$ given $\bar{x}(1),\ldots,\bar{x}(t-1)$ is a normal distribution with mean $\hat{\mu}_p(t)$ and variance

$$\sigma^2 \left\{ \frac{1}{n_t} + \frac{1}{r} \right\}$$

Therefore

$$P[\,\bar{X}(t) \geq ACL_n\,] = 1 - \Phi\left[\frac{ACL_n - \hat{\mu}_p(t)}{\sigma \sqrt{1/n + 1/r}} \right]$$

hence, (22) is equivalent with

$$ACL_n - \hat{\mu}_p(t) \leq -z_{\pi_1} \sigma \sqrt{1/n + 1/r}$$

Substituting ACL_n given by (16) one finds that (22) and (15) are satisfied if

$$RPL + z_{1-\beta}\, \sigma/\sqrt{n} - \hat{\mu}_p(t) + z_{\pi_1} \sigma \sqrt{1/n + 1/r} \leq 0$$

which is equivalent with $\psi_1(n) \geq 0$.

□

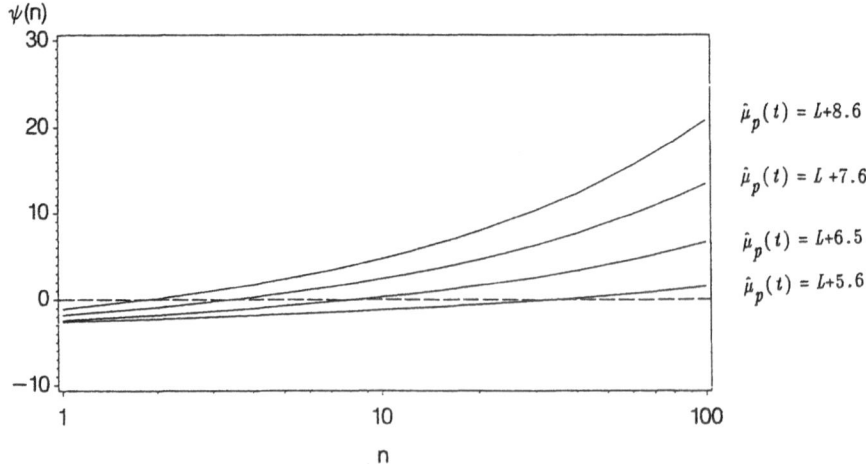

Classical Acceptance Criterion

FIGURE 1: The sample size determining function, $\psi_1(n_t)$ for $\alpha = 0.1$; $\beta = 0.25$; r $= 5$; $\sigma = 3$ and $\hat{\mu}_p(t) = L+5.6$; $L+6.5$; $L+7.6$; $L+8.6$

Thus, a sufficient condition for (15) and (20) to be satisfied is that the sample size, n_t , for the tth subgroup yields a positive value of the sample size determining function $\psi_1(n_t)$. Possible values for n_t are therefore bounded by the roots of $\psi_1(n_t) = 0$, and the condition $n_t > 0$.

Figure 1 shows the variation of $\psi_1(n_t)$ for $\alpha = 0.1$; $\beta = 0.25$; r $= 5.0$; $\sigma = 3$ and selected values of $\hat{\mu}_p(t)$.

6. DETERMINATION OF SAMPLE SIZE UNDER A POSTERIOR ACCEPTANCE CRITERION

It should be noted that the considerations in the previous section used only the process history up to and including subgroup $t\text{--}1$ to determine the *size of the sample* from subgroup t, and that the process history did not otherwise influence the acceptance criterion. In particular, the calculation of the consumer's risk did not utilize the information about the process history that was available prior to the sampling of subgroup t.

However, under the model given by (1) and (2), the previous process history $\bar{x}(1),..,\bar{x}(t\text{--}1)$ also provides information about the current process level $\mu(t)$. Thus, after observation of $\bar{X}(t) = \bar{x}(t)$ one might want to include this information in the assessment of the process output from subgroup t.

In the present section we shall therefore consider an acceptance criterion that is based upon this recursively updated information about current process level $\mu(t)$ as expressed by the conditional distribution of of $\mu(t)$ given the complete process history $\bar{x}(1),..,\bar{x}(t\text{--}1),\bar{x}(t)$.

Following inspection of subgroup t, the information about the current process level is described by the conditional distribution of $\mu(t)$ given the complete process history $\bar{x}(1),..,\bar{x}(t\text{--}1)$, and $\bar{x}(t)$, the posterior distribution. Thus, one may determine the (posterior) probability of an unsatisfactory process level, P$[\mu(t) \leq RPL| \bar{x}(1),..,\bar{x}(t)]$, and use the magnitude of this probability as a basis for the decision whether to accept or reject the output corresponding to subgroup t.

We shall consider the following acceptance criterion

Accept if P$[\mu(t) \leq RPL| \bar{x}(1),..,\bar{x}(t)] \leq \beta$

Reject if P$[\mu(t) \leq RPL| \bar{x}(1),..,\bar{x}(t)] > \beta$

$$(23)$$

The criterion serves to limit the consumer's risk to β, in the sense that the subgroup is only accepted if the (posterior) probability of an unsatisfactory process level does not exceed β.

Although the criterion relates to the distribution of $\mu(t)$ after the sample from subgroup t has become available, the criterion may be formulated in terms of an acceptance control limit that may be determined prior to sampling. The acceptance control limit corresponding to (23) is given in

LEMMA 2

Let the sample size, n_t, for subgroup t be given.

The condition (23) is equivalent to the following acceptance control limit for $\bar{X}(t)$:

$$ACL_n' = RPL + \frac{r}{n_t} \{RPL - \hat{\mu}_p(t)\} + z_{1-\beta} \sigma \sqrt{n_t + r} / n_t \tag{24}$$

with

$$r = 1/\gamma(t)$$

and where $\hat{\mu}_p(t)$ denotes the prediction of $\mu(t)$, and $\gamma(t)$ the corresponding precision parameter determined by successive applications of (5) to (8).

Proof:

Utilizing that the conditional distribution of $\mu(t)$ given the complete process history $\bar{x}(1), .., \bar{x}(t-1), \bar{x}(t)$ is a normal distribution with mean $\hat{\mu}_f(t)$ and variance $\sigma_f^2(t)$ we may reformulate (23) as

Accept if $\hat{\mu}_f(t) \geq RPL + z_{1-\beta} \sigma_f(t)$

Inserting (8) and (11) into the acceptance criterion we obtain the rule:

Accept if $\frac{n_t}{n_t + r} \bar{x}(t) + \frac{r}{n_t + r} \hat{\mu}_p(t) \geq RPL + z_{1-\beta} \sigma/\sqrt{n_t + r}$

Rearranging the terms we obtain (24).

□

Comparing the acceptance control limit ACL_n corresponding to a classical consumer's risk β with the acceptance control limit ACL_n' corresponding to a consumer's posterior risk β we find from (16) and (24)

$$ACL_n' \leq ACL_n \quad \text{for} \quad \hat{\mu}_p(t) \geq RPL + z_{1-\beta} \sigma \{\sqrt{n_t + r} - \sqrt{n_t}\} / r$$

Thus, when the predicted value of $\mu(t)$ is sufficiently far above RPL then the acceptance control limit ACL_n' corresponding to the consumer's posterior risk may be lower than the control limit corresponding to the classical consumer's risk.

Now. consider the determination of the sample size for subgroup t. Since the acceptance criterion already takes the event $\mu(t) \geq RPL$ into consideration. we shall use a simpler requirement than (20). Thus, following inspection of subgroup $t-1$. we want to determine the sample size, n_t , such that

$$P[\ \bar{X}(t) \geq ACL'_n\ ] \geq 1-\alpha \tag{25}$$

is satisfied.

THEOREM 2

Consider the situation following inspection of subgroup $t-1$.

Let $\psi_2(n)$ be given by

$$\psi_2(n) = \kappa \sqrt{r} \sqrt{n+r} - z_{1-\alpha}\sqrt{n} - z_{1-\beta}\sqrt{r} \tag{26}$$

with

$r = 1/\gamma(t)$

$\kappa = \{\ \hat{\mu}_p(t) - RPL\}/\sigma$

where $\hat{\mu}_p(t)$ denotes the prediction of $\mu(t)$ and $\gamma(t)$ the corresponding precision parameter determined by successive applications of (5) to (8).

If the acceptance rule is given by

accept if $P[\ \mu(t) \leq RPL|\ \bar{x}(1),..,\bar{x}(t)\ ] \leq \beta$

and if n_t satisfies

$\psi_2(n_t) \geq 0$

then

$$P[\ \bar{X}(t) \geq ACL'_n\ ] \geq 1-\alpha$$

will hold for $n = n_t$

Proof:

Considering the predictive distribution of $\bar{X}(t)$ given $\bar{x}(1),..,\bar{x}(t-1)$.

We find

$$P[\ \bar{X}(t) \geq ACL'_n\ ] = 1 - \Phi\left[\frac{ACL'_n - \hat{\mu}_p(t)}{\sigma \sqrt{1/n + 1/r}}\right]$$

hence, (25) is equivalent with

$$ACL_n' - \hat{\mu}_p(t) \leq - z_{1-\alpha} \sigma \sqrt{1/n + 1/r}$$

Substituting ACL_n' given by (24) one finds that (23) and (25) are satisfied if

$$\left(1 + \frac{r}{n_t}\right) \{ RPL - \hat{\mu}_p(t) \} + z_{1-\beta}\, \sigma\sqrt{n_t + r}\,/n_t + z_{1-\alpha}\sigma\sqrt{1/n_t + 1/r} \leq 0$$

which is equivalent with

$$\kappa \sqrt{r}\sqrt{n_t + r} - z_{1-\alpha}\sqrt{n_t} - z_{1-\beta}\sqrt{r} \geq 0$$

$\square$

Thus, a sufficient condition for (23) and (25) to be satisfied is that the sample size, n_t, for the tth subgroup yields a positive value of the sample size determining function $\psi_2(n_t)$. Possible values for n_t are therefore bounded by the roots of $\psi_2(n_t) = 0$, and the condition $n_t > 0$.

Figure 2 shows the variation of $\psi_2(n_t)$ for $\alpha = 0.1$; $\beta = 0.25$; r $= 5.0$; $\sigma = 3$ and selected values of $\hat{\mu}_p(t)$.

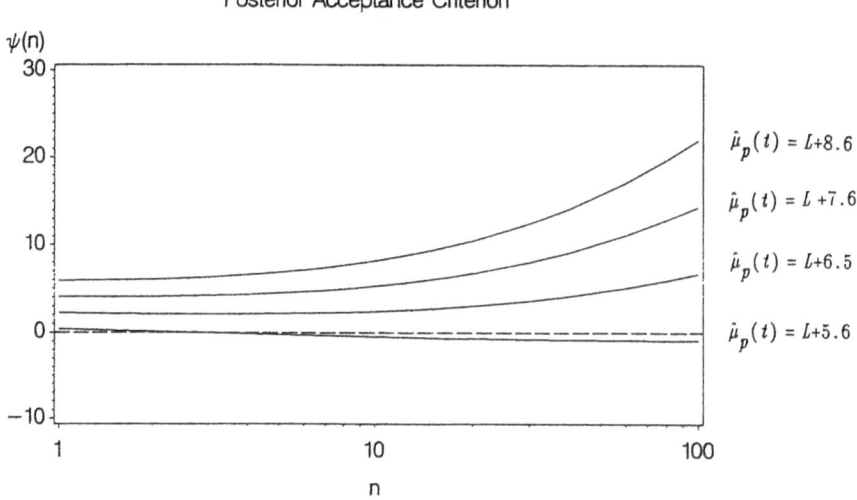

FIGURE 2: The sample size determining function $\psi_2(n_t)$ for $\alpha = 0.1$; $\beta = 0.25$; r $= 5.0$; $\sigma = 3$ and $\hat{\mu}_p(t) = L+5.6$; $L+6.5$; $L+7.6$ and $L+8.6$.

Comparing the graphs in Figure 2 with those in Figure 1, one finds that generally the graph of $\psi_2(n)$ will lie above $\psi_1(n)$ which means that application of the posterior criterion requires a smaller sample size than the classical criterion.

7. DISCUSSION

In the presentation above we have focused upon the application of the acceptance control chart as a means of verifying whether output from a given period is unsatisfactory, or not. This approach does not consider the alternative application of the control charting technique as a means of verifying whether the process in a statistical control, or not.

It may therefore be appropriate to use some charting technique to verify whether the process is in a statistical control. However, since the process model given by (1) and (2) implies that successive subgroup averages are correlated, a direct charting of $\bar{x}(t)$ would be misleading. Instead, one could consider the so—called *innovations* of the process, i.e., the one step ahead forecast errors

$$e(t) = \bar{x}(t) - \hat{\mu}_p(t)$$

and use some charting technique. like an $\bar{X}$–chart or a CUSUM chart to control the sequence of innovations, or squared innovations. Sastri (1988) has investigated the behaviour of a change detection Chi–square variable suggested by Box and Tiao (1976), and Yourstone and Montgomery (1989) have investigated the applicability of geometric moving average, and geometric moving range chart for control of the innovations.

In acceptance control chart applications under a posterior acceptance criterion it is, however of equal importance to make sure that the process model that is used for calculation of the acceptance criterion is still valid. Therefore an adaptive estimation of the process parameters (here: α, σ^2 and σ_0^2) is also appropriate. The papers by Sastri (1988), and Yourstone and Montgomery (1989) above provide some references on this topic, other procedures may be found in the monograph by Ljung and Söderström (1983), or Ljung (1987).

REFERENCES

[1] FREUND, R.A. (1957): Acceptance Control Charts,
   Industrial Quality Control, **XIV**, No. 4. October 1957

[2] BOX, G.E.P. and TIAO, G.C. (1976): Comparison of Forecast and Actuality,
   Applied Statistics **25** : 195–200.

[3] HARVEY, A.C. (1981): *Time Series Models*. New York: Halsted Press.

[4] HUNTER, J.S. (1986): The Exponentially Weighted Moving Average,
   Journal of Quality Technology, **18** : 203–210

[5] JASWINSKI, A.H. (1970): *Stochastic Processes and Filtering Theory.*
   New York: Academic Press.

154

[6] MEINHOLD, M.R. and SINGPURVALLA, N.D. (1983): Understanding the Kalman Filter, *The American Statistician*, **37** : 123–127

[7] LJUNG, L. and SÖDERSTRÖM, T. (1983): *Theory and Practice of Recursive Identification.* Cambridge: MIT Press.

[8] LJUNG, L. (1987): *System Identification: Theory for the User.* New Jersey: Prentice–Hall.

[9] REYNOLDS, M.R. and ARNOLD, J.C. (1989): Optimal One–Sided Shewhart Control Charts With Variable Sampling Intervals, *Sequential Analysis* **8** : 51–57

[10] REYNOLDS, M.R., AMIN, R.W., ARNOLD, J.C. and NACHLAS, J.A. (1988): X̄ Charts With Variable Sampling Intervals, *Technometrics* **30** : 181–192.

[11] SASTRI, T. (1988): An Adaptive Estimation Algorithm, *IIE Transactions* **22** : 176–184

[12] YOURSTONE, S.A. and MONTGOMERY, D.C. (1989): A Time–Series Approach to Discrete Real–Time Process Quality Control, *Quality and Reliability International*, **5** : 309–317.

The Use (and Misuse) of False Alarm Probabilities in Control Chart Design

B.M. Adams and W.H. Woodall, Tuscaloosa, Alabama, USA
C.A. Lowry, Fort Worth, Texas, USA

1. Introduction

An increased awareness in recent years of the impact of quality on the marketplace has renewed interest in the field of statistical process control. The result of this interest has been an increase in the development and analysis of process monitoring schemes. In this paper we discuss certain methods of designing control charts for the process mean based upon the probability of incorrectly concluding a shift in the process mean has occurred and adjustment is necessary, i.e. the probability of a Type I error, α.

Two types of charts will be examined, the standard Shewhart chart augmented with runs rules and the cumulative sum (CUSUM) control chart. Our primary focus will be on ambiguous definitions of α, the relationship between α and the relative frequency of false alarms, and the accuracy with which the design methods result in monitoring schemes reflecting the desired characteristics.

Section 2 begins with a discussion of the standard Shewhart chart and the Shewhart chart augmented with runs rules. The definition and computation of α along with its relationship to the relative frequency of false alarms is examined. The CUSUM chart and the design method of Johnson (1961) for obtaining CUSUM schemes are discussed in Sections 3 and 4. A table of actual versus desired in-control average run lengths, displaying the poor performance of Johnson's method of CUSUM chart design, is presented in Section 5. The theoretical foundation for Johnson's (1961) method of designing CUSUM charts is reviewed in Section 6. A clear definition of α and its relationship to the average run length of the CUSUM chart is provided.

2. The Shewhart Chart

The basic Shewhart charts have found many useful applications since the 1931 release of Dr. Shewhart's book, *Economic Control of Quality of Manufactured Product*. The $\overline{X}$-chart is used to monitor the mean of a quality characteristic for a given process. Observations are taken periodically and the sample mean is plotted. If we let X_{ij} represent the jth measurement in the ith sample, $i = 1,2,...,m$ and $j = 1,2,...,n$, then the ith sample mean is given by

$$\overline{X}_i = (\sum_{j=1}^{n} X_{ij})/n.$$

Frontiers in Statistical Quality Control 4
Ed. by Lenz et al.
© Physica-Verlag Heidelberg 1992

It is assumed that the sample means are mutually independent and that $\overline{X}_i \sim N(\mu, \sigma^2/n)$. A signal occurs on the $\overline{X}$ - chart if $\overline{X}_i$ is more than $3\sigma/\sqrt{n}$ units away from the in-control mean, μ_0. The occurrence of a signal when the mean of the quality characteristic is on target is referred to as a false alarm. In the context of hypothesis testing a false alarm would be referred to as a Type I error. Generally μ and σ^2 are unknown and are estimated by using historical data. The reader is referred to Wadsworth (1990), Montgomery (1985) or other standard texts for estimating procedures. In our paper we consider, for simplicity, the performance of control charts to monitor the process mean under the assumption that the process variance is known and remains stable.

The assumption of mutually independent samples allows one to compute meaningfully the probability, α, of an observation producing a false alarm on the $\overline{X}$-chart. The probability of a false alarm is constant over time and does not depend on observations from previous time periods. The run length of a control procedure is the number of observations made until a signal occurs. For the case of the standard $\overline{X}$ - chart, the run length has a geometric distribution. Hence, the expected number of observations until a signal occurs, also known as the average run length (ARL), given that the process is on target, is related to α by the well-known formula

$$ARL = 1/\alpha \, , \tag{1.1}$$

where $\alpha = P\,(|\overline{X}_i - \mu_0| > 3\sigma/\sqrt{n})$. Equation (1.1) provides an intuitively pleasing relationship between the false alarm probability and the frequency of false alarms for the in-control state. The α-value of .0027, corresponding to the 3-sigma limits, would imply a false alarm rate of (on average) 1 in every 370 observations given that the process is on target. Equivalently, there will be 370 observations, on the average, between false alarms.

The basic Shewhart chart can be augmented by other rules, some of which are referred to as runs rules, for identifying unusual patterns in the points on a control chart. Commonly applied runs rules which are useful in detecting a small sustained shift in the mean include the following:

i)   Two out of three consecutive points on the same side of the center line and beyond two standard errors from the mean.

ii)   Four out of five consecutive points on the same side of the center line and more than one standard error from the mean.

iii)   Eight consecutive points on the same side of the center line.

Additional runs rules and information on the interpretation of control chart patterns can be found in the Western Electric *Statistical Quality Control Handbook* (1956, pp. 149 - 180) or Nelson (1985).

In situations where runs rules are incorporated into the standard Shewhart chart, the determination of a false alarm probability is more difficult. The value of α is no longer constant over time because the probability of a false alarm at a given observation is dependent upon the values of previous observations. Montgomery (1985, p. 115), Banks (1989, p. 141), and Duncan (1986, p. 435) provide an approximation of α when k runs rules are used in a control scheme and each rule m has a Type I error probability of α_m, $m=1,2,...k$. The overall Type I error probability for the control scheme based on all k tests is approximated by

$$\alpha = 1 - \prod_{m=1}^{k} (1 - \alpha_m). \tag{2.1}$$

As Montgomery (1985) and Duncan (1986) state, this approximation oversimplifies the problem of approximating the overall Type I error probability by assuming the independence of the k runs rules. In addition to this oversimplification, the value α_m itself is usually not well-defined since the probability of a false alarm by rule m at any given time point is not constant but depends on the values of past observations.

As an example, consider the standard Shewhart chart with the 3 supplementary runs rules listed above. Figure 1 displays the regions of the Shewhart chart defined by these runs rules and the respective in-control probabilities of a sample mean falling in each region. Let α_i denote the probability of a signal resulting from runs rule (i). The Western Electric *Statistical Quality Control Handbook* (1956) and Grant and Leavenworth (1980) determine the probability of two out of three consecutive points falling in region A_1 or beyond using the binomial formula. Ott (1975, p. 49) and Nelson (1985) use a similar approach. Letting X equal the number of sample means falling in region A_1 or beyond and assuming X has a binomial distribution with $n=3$ and $p=.0227$, they obtain $\alpha_1 = .0015$. The same value is obtained for the probability of a signal resulting from two of three consecutive points falling in region A_2 or beyond. Proceeding in the same manner for the remaining rules and using equation (2.1) results in an estimate of $\alpha = .01865$. The Western Electric handbook estimates the overall probability of a Type I error by $\alpha = \sum \alpha_m$, which gives a very similar value. The value $\alpha = .01865$ would imply an in-control ARL ($= 1/\alpha$) of 53.62. Champ and Woodall (1987), however, obtain an exact in-control ARL of 91.75 which is over 70% higher than the estimate provided by equation (2.1).

Many questions concerning the use of equation (2.1) arise from the above example. A few questions, for instance, are

i) In computing α_1 should one include the probability of falling beyond 3 standard errors above the mean?

ii) Should the rule which signals if 2 of 3 consecutive points are in region A_1 or beyond be considered a separate rule from that based on 2 of 3 consecutive points in region A_2 or beyond? (Obviously the two rules do not operate independently.)

iii) Should conditional probabilities be used in the binomial formula? (If so, conditioned on what?)

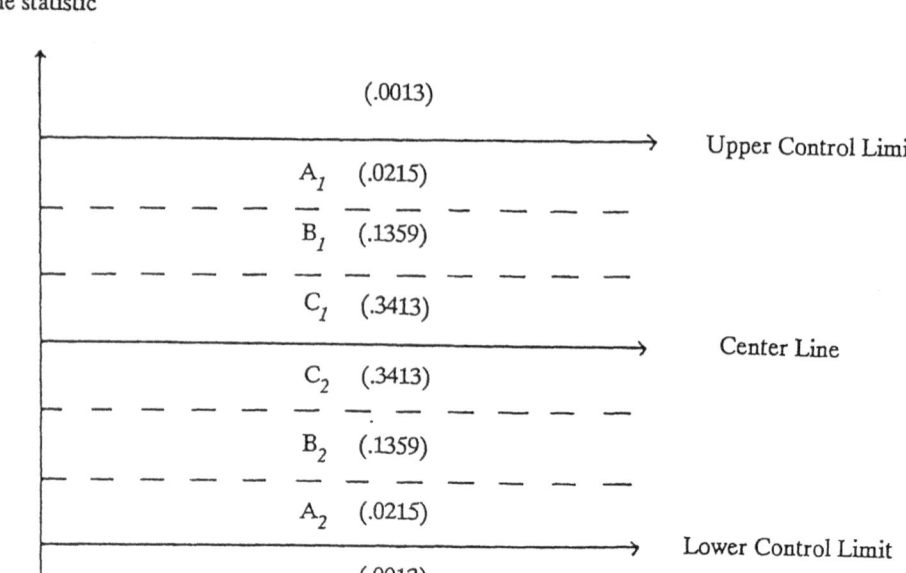

Value of
the statistic

(.0013)

Upper Control Limit

A₁ (.0215)

B₁ (.1359)

C₁ (.3413)

Center Line

C₂ (.3413)

B₂ (.1359)

A₂ (.0215)

Lower Control Limit

(.0013)

Sample Number

Figure 1. The Shewhart $\overline{X}$-Chart with Runs Rules

Unfortunately, the rather vague method for approximating α using equation (2.1) answers none of these questions. Hence, the probability of a false alarm for the Shewhart chart augmented with runs rules is ill-defined in any practical sense. With these difficulties in mind, defining α as "the probability of a false alarm" can lead to confusion. It is unclear whether one is speaking of the probability of a false alarm for a given observation (which should vary depending on previous observations) or of the relative frequency of false alarms.

Defining α as the relative frequency of false alarms allows one to design control procedures based on desired in-control average run lengths. Champ and Woodall (1987, 1990) give an exact approach for determining ARL's for the Shewhart chart combined with runs rules. Their Markov chain approach models the dependencies between the rules and allows for the fact that runs rules are based on past observations.

3. The Cumulative Sum Control Chart

The cumulative sum (CUSUM) control chart has received much attention in quality control literature. The developmental work of Page (1954, 1961) and Barnard (1959) has been built upon by Johnson (1961), Johnson and Leone (1962a, 1962b, 1962c), Lucas (1973, 1982), Lucas and Crosier (1982), Woodall (1983, 1984), and many others. Much of this work is summarized by Goel (1982) and Woodall (1986). The CUSUM chart has become a standard topic in introductory quality control texts such as Montgomery (1985), Wadsworth, Stephens, and Godfrey (1986), Duncan (1986) and others. Many SPC software packages such as Quality Alert (V 4.0), SPC-PCII, and SAS/QC now support CUSUM applications. These texts and software packages obtain the defining values of the V-mask representation of the CUSUM chart by using two parameters, the probability of a false alarm, α, and the size of the smallest shift, δ, one wishes to detect quickly. We show in Sections 4-6 that the value α often used in the chart's design is misinterpreted and misused.

The CUSUM chart can be represented by a V-mask scheme or by an equivalent two-sided decision interval scheme. After the ith sample, the value of the CUSUM statistic for the V-mask is

$$ S_i^{'} = \sum_{j=1}^{i} (X_j - \mu_0)/\sigma_X, \quad i = 1,2,3,\dots . $$

Hence, the CUSUM statistic is a cumulative sum of the deviations of the observed sample means from the target value. The cumulative sums are plotted against time. A signal occurs if the sum increases or decreases so rapidly that a point falls outside the boundaries defined by the CUSUM V-mask.

To use the V-mask representation, the point A of the V-mask, as shown in Figure 2, is placed on the most recently plotted point of the CUSUM chart. A shift in the process mean is indicated by any point falling outside an arm of the V-mask. A point falling below the lower arm of the V-mask would indicate an increase in the process mean, while a point falling above the the upper arm of the V-mask would indicate a decrease in the process mean. It is not necessarily the point falling outside the arms of the V-mask which is suspect, but the pattern of increases or decreases in the subsequent points. The slope of the points in the increasing or decreasing pattern measures the average deviation from the value of μ_0.

Figure 2 illustrates the CUSUM chart represented by a V-mask scheme. These statistics were computer-generated using samples of size $n = 4$. The first twenty samples were from the $N(0, 1)$ distribution while the last seven samples were from the $N(0.5, 1)$ distribution. A signal occurs at the 27th observation as the 21st point falls outside the lower arm of the V-mask. One may note the upward trend in the cumulative sum points beginning around the 20th time period.

The V-mask representation of the CUSUM chart has two major disadvantages. The first disadvantage is a tendency of the cumulative sums to wander considerably (perhaps off a sheet of graph paper) even if the process mean is on target. Secondly, it is not clear how many of the previous values of the cumulative sum must be remembered for the signaling scheme.

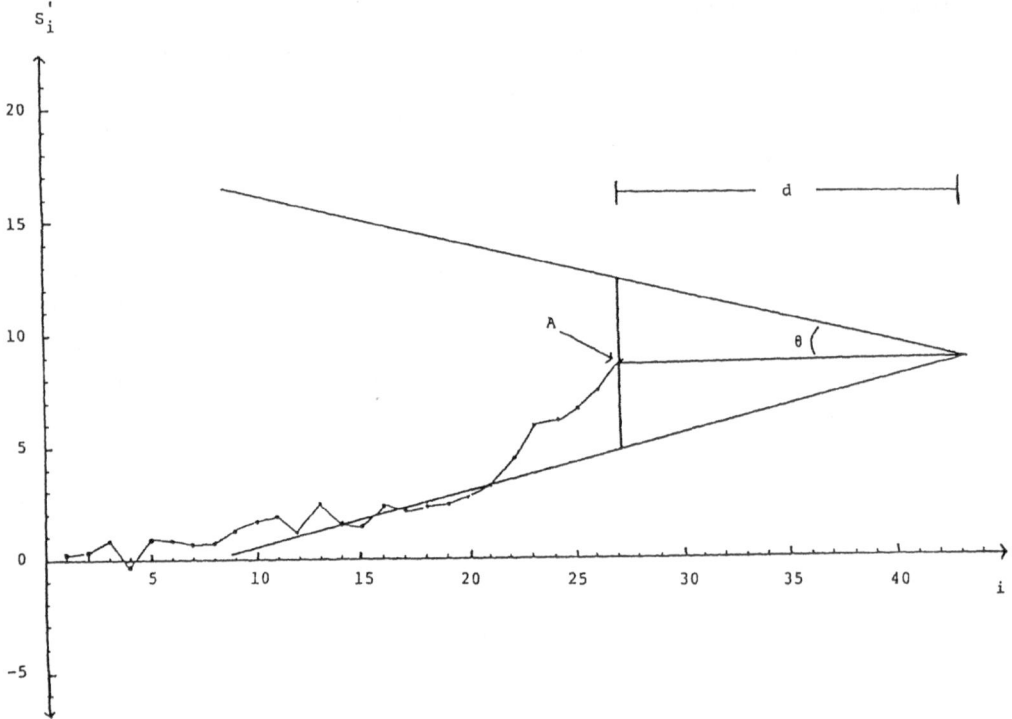

Figure 2.   The CUSUM Chart with V-mask

A mathematically equivalent alternative to the CUSUM chart with a V-mask, which overcomes these disadvantages, is the two-sided decision interval CUSUM chart. Two cumulative sums are defined as

$$S_i = \max(0, S_{i-1} + Z_i - k)$$

and

$$T_i = \min(0, T_{i-1} + Z_i + k)$$

where $Z_i = (\bar{X}_i - \mu_0)/\sigma_{\bar{X}}$, $i = 1,2,3,....$ Typically, the initial values are $S_0 = T_0 = 0$, and the reference value k is $\delta/2$. These sums are plotted against time and a signal occurs if $S_i > h$ or $T_i < -h$ for a given constant h > 0.

Figure 3 illustrates the two-sided decision interval CUSUM which corresponds to the V-mask representation given in Figure 2. The data used in Figure 3 is also the same as that used to produce Figure 2.

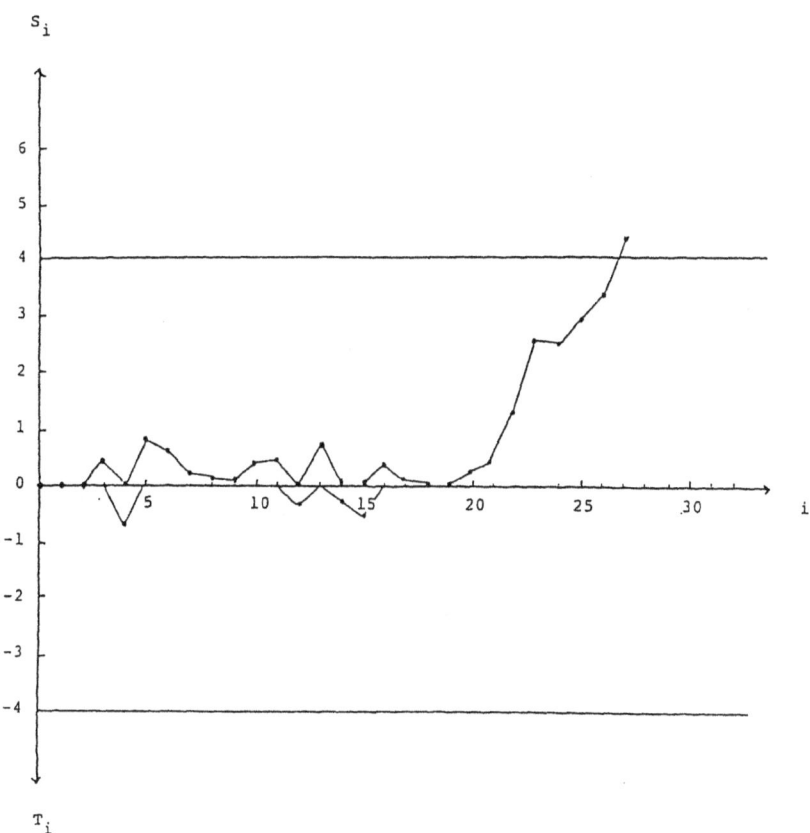

Figure 3. The Two-sided Decision Interval CUSUM Chart

4. The Design of Cusum Charts

The design of the CUSUM chart consists of determining the value of the lead distance d, and the angle θ, which define the V-mask. The determination of these two values is often based on the size of the shift one wishes to detect quickly, denoted by δ, and the probability of a false alarm, denoted by α. The following guidelines for obtaining d and θ were given by Johnson (1961):

i) Determine the smallest shift, δ, in the mean which one desires to detect quickly (where δ is measured in units of $\sigma/\sqrt{n}$.)

ii) Determine the greatest tolerable probability of a false alarm (a signal when the process is on target). Denote this probability by 2α.

iii) Let
$$\theta = \arctan(\delta/2) \tag{4.1}$$
and
$$d = -2\,\delta^{-2}\ln(\alpha). \tag{4.2}$$

These guidelines are also given by Montgomery (1985, p. 228), Wadsworth et al. (1986, p. 338), Goel (1982, p. 239), Wetherill (1977), and Puri (1984, p. 203), among others. This approach is also used in the SPC software packages mentioned in Section 3. Johnson's approach is modified slightly to design CUSUM charts for other applications, such as monitoring process variability, by Johnson and Leone (1962a, 1962b, 1962c) and Johnson and Leone (1964).

The design for the two-sided decision interval CUSUM chart can be obtained directly from the V-mask using the relationships derived by Kemp (1961),

$$k = w \tan(\theta)$$

and

$$h = d \tan(\theta),$$

where w equals the horizontal distance between consecutive points on the V-mask in terms of unit distance on the vertical scale. The performance of the CUSUM scheme designed using equations (4.1) and (4.2) will be discussed in the next section.

5. On Target CUSUM ARL's

The actual in-control ARL's for CUSUM charts obtained using equations (4.1) and (4.2) are compared with the desired ARL ($=1/2\alpha$) in Table 1. This comparison is important for several reasons. One reason is to determine whether the design method does or does not actually provide a control scheme with the desired properties. Another reason is that performance comparisons among charting methods is often based on schemes with equal in-control ARL's. Hence, fair comparisons rely on accurate designs.

Table 1 provides the actual and desired ARL's for CUSUM charts designed using Johnson's method. Various levels of δ and α are provided. The actual ARL's were obtained by using the numerical integration techniques of Woodall (1983). As an example, the table shows for $\alpha = 0.005$ ($2\alpha = 0.01$ for two sided chart) and $\delta = 1$ Johnson's method leads to a control scheme with an in-control average run length of 629.44. This value is substantially different from the desired ARL of 100 implied by the use of the 2α-value 0.01 and the relationship ARL = $1/2\alpha$.

	α (one-tail) (ARL = 1/2α)					
δ	0.05 (10)	.025 (20)	.01 (50)	.005 (100)	.00135 (370)	.001 (500)
1.0	58.55	121.5	311.6	629.4	2350.6	3184.4
1.2	51.72	106.9	273.0	550.0	2050.3	2769.4
1.4	48.07	99.1	252.2	508.1	1891.1	2553.0
1.6	46.23	95.1	242.2	486.9	1812.0	2445.0
1.8	45.85	93.5	238.6	480.6	1787.6	2414.4
2.0	46.74	94.1	'239.9	484.0	1804.5	2435.8
2.2	48.78	96.8	244.0	493.4	1848.0	2499.8
2.4	51.97	101.3	252.2	509.0	1915.4	2591.1
2.6	56.57	108.2	263.9	529.2	1994.1	2706.8
2.8	62.78	117.4	281.1	556.0	2087.5	2830.7
3.0	70.90	130.2	303.6	592.6	2194.8	2975.4

Table 1.   In-Control ARL's for the Two-sided CUSUM Chart

Table 1 shows that the actual in-control average run lengths are generally much greater than implied by the α-value used in Johnson's approach to determine the control procedure. In other words, Table 1 illustrates the inaccuracy of the relationship ARL $= 1/2\alpha$ with Johnson's use of α.

6. The Interpretation of α

The difference in the desired ARL's and the actual ARL's for the in-control case is the result of an ambiguous use of α. Many authors recommending Johnson's procedure define α in a manner similar to that of Montgomery (1985, p. 228), i.e. as the probability of a false alarm. Unlike the Shewhart chart, the meaning of this definition is not clear for the CUSUM chart. Shewhart charts begin with assumptions which include independence within and between samples. Independence between samples allows one to designate a false alarm probability which is constant from observation to observation. While independence within samples is assumed for the CUSUM chart, consecutive CUSUM statistic values are clearly not independent. Defining α in terms of a false alarm probability at a given observation becomes confusing as that probability changes from observation to observation. Hence, it seems unreasonable to assume this definition of the probability of a false alarm.

Some authors, such as Bicking and Gryna (1979), incorrectly suggest that the CUSUM chart determined using Johnson's method for $\alpha = 0.00135$ has limits comparable with the 3-sigma limits used in the Shewhart control chart. The Shewhart $\bar{X}$-chart with 3-sigma limits has an in-control ARL of 370.4. From Table 1 it can be seen that the CUSUM charts designed using Johnson's method with $\alpha = .00135$ have in-control ARL's of roughly 2000.

Let us consider the meaning of α in the development of Johnson's method. Johnson (1961) states that the CUSUM chart is roughly equivalent to a sequential probability ratio test (SPRT) "in reverse." Figure 4 is a diagram of an SPRT for distinguishing between the three hypotheses for the mean of the Z_i's, $H_{-1}: \mu = -\delta$, $H_0: \mu = 0$, and $H_1: \mu = \delta$. The critical feature of the diagram is that the plane is divided into 4 action regions: accept H_1, accept H_{-1}, accept H_0, or continue sampling. The sequence of values

$$Y_1 = Z_m, \quad Y_2 = Z_{m-1}, \quad \text{and so on}$$

are calculated and the values of the cumulative statistic $S_i'' = \sum_{j=1}^{i} Y_j$ are plotted against time. Note that the Y_i's are in reverse time order from the X_i's. In the case of an SPRT, sampling continues until an observation falls into a region in which a hypothesis is accepted and sampling ends. Johnson (1961) shows that the probability of falsely rejecting H_0, α, for this procedure is related to the distance d by

$$d \approx 2\,\delta^{-2}\ln((1-\beta)/\,\alpha),$$

where β is the probability of a Type II error. Johnson (1961) states for small β, $\ln(1-\beta) \approx 0$ and $d \approx -2\delta^{-2}\ln(\alpha)$, giving the expression in equation (4.2).

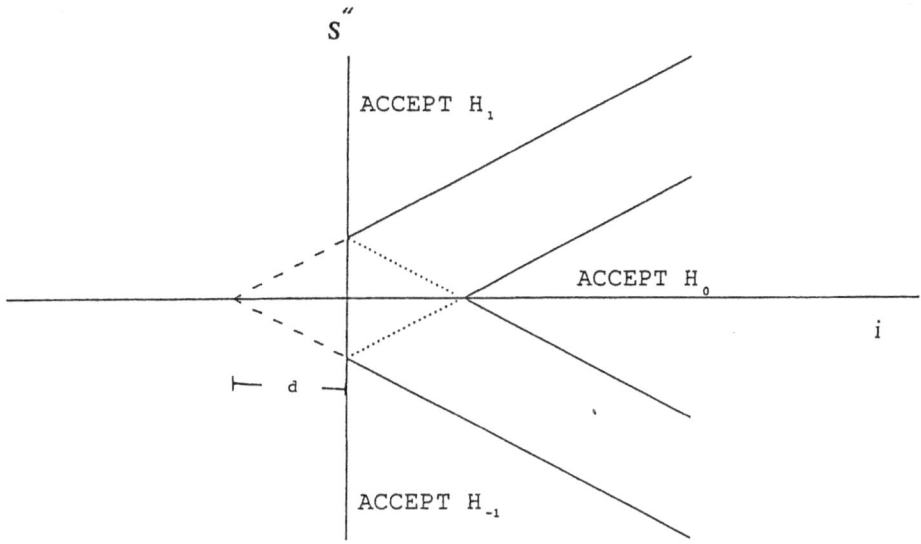

Figure 4. The Two-Sided Sequential Probability Ratio Test

The procedure based on the values of the Y_i's defined above results in a decision scheme which is neither an SPRT nor a CUSUM chart. This is evident from the fact that the entire sequence of plotted values changes with each observation. A proper relationship between the SPRT and the CUSUM chart (for the one-sided case) is given by Ewan and Kemp (1960). They show that the one-sided CUSUM chart is equivalent to a sequence of one-sided SPRT's. The two-sided CUSUM chart is not, however, equivalent to a sequence of two-sided SPRT's.

The CUSUM chart, although similar in appearance to the SPRT in reverse, has no mechanism for accepting the null hypothesis. The result of this difference is an ambiguity in the meaning of α. As observations are made on a process which is on target, the probability of eventually obtaining a false alarm for the CUSUM chart must be equal to one since there is no formal criteria for accepting the hypothesis that the process is on target. Hence, defining α as the overall probability of a false alarm is meaningless for the CUSUM chart for $\alpha \neq 1$. Similarly, defining β as the overall probability of a Type II error is meaningless for $\beta \neq 0$. Johnson (1961) mentions the structural differences in the CUSUM chart and the backwards SPRT, but the effect of this difference has been ignored. Table 1 shows that this omission is unfortunate.

Faced with this difficulty of interpretation, a meaningful alternative is to define α as the proportion of observations which are false alarms. This is the definition used by van Dobben de Bruyn (1968, p. 18). This interpretation leads to the relationship $\alpha = 1/\text{ARL}$ ($2\alpha = 1/\text{ARL}$ for the two-sided chart). It is this interpretation of α which is used to evaluate Johnson's method in Table 1. Unfortunately, this definition is inconsistent with the definition of α in Johnson's development of the

CUSUM chart design. Johnson states that his α-value is the probability of a false alarm for the complete CUSUM procedure, but as shown above, this interpretation is not reasonable.

The differences in desired and actual ARL's displayed in Table 1 result not only from inconsistencies in defining α, but from the structural differences in the CUSUM chart and the SPRT. For the SPRT, let the expected number of observations given that H_0 is correctly accepted be denoted by $E_0(N)$, let the expected number of observations given a false alarm be denoted by $E_A(N)$, and let the true probability of a Type I error for the SPRT be denoted by α_S. Note that the ARL of the SPRT for the in-control case would be $(1-\alpha_S) E_0(N) + \alpha_S E_A(N)$, i.e., a weighted average of the expected number of observations given a false alarm and the expected number of observations given H_0 is accepted. The CUSUM chart, having no formal acceptance mechanism for H_0, accumulates sequences of observations which would result in the acceptance of H_0 for the SPRT. Hence, the ARL for the CUSUM chart would be closer to $E_A(N) + [(1-\alpha_S) / \alpha_S] E_0(N)$. This relationship would hold if the two-sided CUSUM chart were equivalent to a sequence of two-sided SPRT's.

7. Conclusions

We have pointed out some common misinterpretations and the consequent misuses of false alarm probabilities in the design of control charts. Most of the approaches to approximating the overall false alarm probability for the Shewhart $\overline{X}$ - chart with supplementary runs rules are not accurate. For this type of chart we recommend the Markov chain approach given by Champ and Woodall (1987).

Johnson's (1961) approach to the design of CUSUM charts is based on a false alarm probability that is not carefully defined. A definition of α as the proportion of samples which result in false alarms is presented as the only meaningful definition. We have shown that the in-control ARL's of CUSUM charts determined by Johnson's (1961) method are much higher than those desired based on this definition of α. Despite its inaccuracy, Johnson's approach has been, and continues to be, used in variety of applications. See, for example, Johnson (1963, 1966), Rai (1966), Alwan (1986), and Ramirez (1989).

We recommend that CUSUM charts be designed based on desired ARL properties. Specifically, one can let $k = \delta/2$ and then determine h so that the in-control ARL is some specified value. An in-control ARL of ARL_0 is related to the proportion of samples giving false alarms, α_0, by $\alpha_0 = 1/ARL_0$. Goel and Wu (1971) provide a contour nomogram for CUSUM charts used to control the mean and Vance (1986) provides a computer program. The ARL approach is standard in many textbooks such as Montgomery (1985), Wadsworth et al. (1986), and Banks (1989).

8. References

(1)   Alwan, L.C. (1986), "Cusum Quality Control -Multivariate Approach," *Communications in Statistics - Theory and Methods*, 15 (12), 3531-3543.

(2)   Banks, J. (1989), *Principles of Quality Control*, New York: John Wiley & Sons, Inc.

(3)   Barnard, G.A. (1959), "Control Charts and Stochastic Processes," *Journal of the Royal Statistical Society*, B, 11, 239-271.

(4)   Bicking, C.A. and Gryna, F.M., Jr. (1979), *Quality Control Handbook*, ed. Juran, J.M., Gryna, F.M., Jr., and Bingham, R.S., Jr., New York: McGraw-Hill Book Company.

(5)   Champ, C.W. and Woodall, W.H. (1987), "Exact Results for Shewhart Control Charts with Supplementary Runs Rules," *Technometrics*, 29, 393 - 399.

(6)   Champ, C.W. and Woodall, W.H. (1990), "A Program to Evaluate the Run Length Distributions of a Shewhart Control Chart with Supplementary Runs Rules," *Journal of Quality Technology*, 22, 68-73.

(7)   Duncan, A.J. (1986), *Quality Control and Industrial Statistics*, 5th Ed., Homewood, IL: Richard D. Irwin.

(8)   Ewan, W.D. and Kemp, K.W. (1960), "Sampling Inspection of Continuous Processes with No Autocorrelation between Successive Results," *Biometrika*, 47, 363-380.

(9)   Goel, A.L. (1982), "Cumulative Sum Control Charts," in *Encyclopedia of Statistical Sciences*, Vol. 2, Kotz, S. and Johnson, N.L., editors.

(10)   Goel, A.L. and Wu, S.M. (1971), "Determination of A.R.L. and a Contour Nomogram for Cusum Charts to Control Normal Mean," *Technometrics*, 13, 221 - 230.

(11)   Grant, E.L. and Leavenworth, R.S. (1980), *Statistical Quality Control*, New York: McGraw-Hill Book Company.

(12)   Johnson, N.L. (1961), "A Simple Theoretical Approach to Cumulative Sum Control Charts," *Journal of the American Statistical Association*, 56, 835 - 840.

(13)   Johnson, N.L. (1963), "Cumulative Sum Control Charts for the Folded Normal Distribution," *Technometrics*, 5, 451-458.

(14)   Johnson, N.L. (1966), "Cumulative Sum Control Charts and the Weibull Distribution", *Technometrics*, 8, 481-491.

(15)   Johnson, N.L. and Leone, F.C. (1962a), "Cumulative Sum Control Charts, Mathematical Principles Applied to their Construction and Use, Part I," *Industrial Quality Control*, June, 18, 15-21.

(16)   Johnson, N.L. and Leone, F.C. (1962b), "Cumulative Sum Control Charts, Mathematical Principles Applied to their Construction and Use, Part II," *Industrial Quality Control*, July, 18, 29-36.

(17)   Johnson, N.L. and Leone, F.C. (1962c), "Cumulative Sum Control Charts, Mathematical Principles Applied to their Construction and Use, Part III," *Industrial Quality Control*, August, 18, 22-28.

(18)   Johnson, N.L. and Leone, F.C. (1964), *Statistics and Experimental Design in Engineering and the Physical Sciences*, Vol. 1, New York: John Wiley & Sons, Inc.

(19)   Kemp, K.W. (1961), "The Average Run Length of the Cumulative Sum Chart when a V-Mask is Used," *Journal of the Royal Statistical Society*, B, 23, 149-153.

(20)   Lucas, J.M. (1973), "A Modified 'V' Mask Control Scheme," *Technometrics*, 15, 4, 833-847.

(21)   Lucas, J.M. (1982), "Combined Shewhart-CUSUM Quality Control Schemes," *Journal of Quality Technology*, 14, 51-59.

(22) Lucas, J.M. and Crosier, R.B. (1982), "Fast Initial Response for CUSUM Quality-Control Schemes: Give Your CUSUM a Head Start," *Technometrics*, 24, 3, 199-205.

(23) Montgomery, D.C. (1985), *Introduction to Statistical Quality Control*, New York: John Wiley & Sons, Inc.

(24) Nelson, L.S. (1985), "Interpreting Shewhart $\overline{X}$ Control Charts", *Journal of Quality Technology*, 17, 114-116.

(25) Ott, E.R. (1975), *Process Quality Control*, New York: McGraw-Hill Book Company.

(26) Page, E.S. (1954), "Continuous Inspection Schemes," *Biometrika*, 41, 100-115.

(27) Page, E.S. (1961), "Cumulative Sum Charts," *Technometrics*, 3, 1, 1-9.

(28) Puri, S.C. (1984), *Statistical Process Quality Control*, Ottawa: Standards - Quality Management Group.

(29) Rai, G. (1966), "Cumulative Sum Control Charts for Truncated Normal Distributions," *Australian Journal of Statistics*, 8, 80 - 86.

(30) Ramirez, J.G. (1989), *Sequential Methods in Statistical Process Monitoring*, unpublished Ph.D. dissertation, University of Wisconsin - Madison.

(31) Shewhart, W.A. (1931), *Economic Control of Quality of Manufactured Product*, New York: D. Van Nostrand Co., Inc. (Available from ASQC, 310 Wisconsin Ave., Milwaukee, WI 53203.)

(32) Van Dobben de Bruyn, C.S. (1968), *Cumulative Sum Tests - Theory and Practice*, New York: Hafner Publishing Company.

(33) Vance, L.C. (1986), "Average Run Lengths of Cumulative Sum Control Charts for Controlling Normal Means," *Journal of Quality Technology*, 18, 189-193.

(34) Wadsworth, H.M., Stephens, K.S., and Godfrey, A.B. (1986), *Modern Methods for Quality Control and Improvement*, New York: John Wiley & Sons, Inc.

(35) Wadsworth, H.M. (1990), *Handbook of Statistical Methods for Engineers & Scientists*, New York: McGraw-Hill Book Company.

(36) Western Electric (1956), *Statistical Quality Control Handbook*, Select Code 700-444. (Available from AT&T, P.O. Box 19901, Indianapolis, IN 46219.)

(37) Wetherill, G.B. (1977), *Sampling Inspection and Quality Control*, London: Chapman and Hall.

(38) Woodall, W.H. (1983), "The Distribution of the Run Length of One-Sided CUSUM Procedures for Continuous Random Variables," *Technometrics*, 25, 295-301.

(39) Woodall, W.H. (1984), "On the Markov Chain Approach to the Two-sided CUSUM Procedure," *Technometrics*, 26, 41-46.

(40) Woodall, W.H. (1986), "The Design of CUSUM Quality Control Charts," *Journal of Quality Technology*, 18, 2, 99-102.

Optimal Dynamic Application of a Fraction-Defective Control Chart to Control a Production Process

S. P. Ladany and T. Raz, Beer Sheva, Israel

1. INTRODUCTION

When a manufacturing process is controlled with a Fraction-Defective Control-Chart, the values of 3 decision variables have to be determined: the timing of the samples, the size of the samples, and the distance of the control limits from the central line on the control-chart in terms of standard deviations.

Various authors suggested different models to determine when to inspect a statistically failing single system. Starting with the classical work of Barlow et al. [2] and Barlow & Proschan [3], the literature includes a variety of specialized models, such as those of Wattanapanom & Shaw [23] for inspection that affect the failure rate, Sackrowitz & Samuel-Cahn [20] and Luss [14] for Markovian deterioration, Keller [10] for numerous inexpensive inspections, Luss & Kander [15] for inspection of non-negligible durations, Mumford [18] for increasing hazard rates, Sengupta [22] for delayed failure rates, and Luss [14] for production facilities.

Most of the research, Kuss & Kander [15] being one exception, ignored the occurrence of inspection errors of type I (classification of an acceptable item or process as unacceptable) and of type II (classification of an unacceptable item or process as acceptable). Raz & Ladany [19] incorporated such errors in their optimal discrete time inspection policy using a Markov chain model of the system state, and as such they determined equal-sample-size equally-spaced steady-state policies.

While Luss & Kander [15] and Luss [14] dealt with transient-period inspection policies using Dynamic Programming, their results were not applied to the framework of a production process controlled with control-charts. On the other hand, within the framework of optimal economic design of quality control procedures (reviewed by Montgomery [16]), where Duncan [6] and Gibra [7], Knappenberger & Grandage [11], Baker [1], Chiu & Wetherill [5], Goel [8], and Hall & Eilon [9], Saniga & Montgomery [21] dealt with $\bar{X}$-charts, Montgomery & Klatt [17] dealt with multivariate extensions of the $\bar{X}$-chart, and Ladany [12], Chiu [4], and Ladany & Bedi [13] dealt with fraction-defective control charts, all the models derived steady-state equal-sample-size equally-spaced policies.

For a production process which is <u>perfectly</u> set up at the start of each working shift, and which might be only in one of 2 possible states (generating p_1 or p_2 fraction of defectives), the optimal values of the above stated decision variables have been determined by Ladany [12], and the results have been used by Ladany & Bedi [13] for the selection between scheduled and unscheduled set-ups at the end of work-shifts. These works determined those optimal values for the equal-policy case in which samples of the same size are taken at constant time intervals. The optimal values were found for the objective function of minimizing total cost.

Frontiers in Statistical Quality Control 4
Ed. by Lenz et al.
© Physica-Verlag Heidelberg 1992

However, the fact is that if the process is perfectly set up at its outset, and the change from one state to another is random, then it is logical to expect that the sequence of samples should be spaced at decreasing time intervals from each other, and/or should have increased sample sizes. This should be envisaged during the transient period, which is characterized by the fact that the probabilities — that the process is in a given state — decrease (or increase) with the passage of time and with the accumulation of sampling errors since the process' initial perfect set up.

Thus, the aim of this paper is to find the Dynamic Optimal Sampling policy which is not constrained by a requirement of "equal sampling interval policy." Though theoretically possible, it does not seem to be practical (from usage point of view) to consider also changes from sample to sample in the distance of the control limits from the central line in terms of standard deviations. Nevertheless, it will be accounted for in the general formulation of the problem, though it is envisaged that practical solutions will call for the adoption of the common practice of using control charts at the distance of 3 (or any other number of) standard deviations from the central line. In addition, the objective function accommodates managements' aim to maximize profit, rather than to simplify the calculations and to suffer sub-optimal results under a cost minimization approach.

Section 2 outlines the probabilistic framework of the problem. Section 3 describes the Dynamic Optimization Model, Section 4 deals with the solution procedure, while Section 5 provides a numerical example.

2. THE PROBABILISTIC FRAMEWORK

The process which is perfectly set up at the beginning of the shift, generates p_1 fraction of defectives. The process might jump randomly from the state of p_1, to a state in which it produces p_2 fraction of defectives ($p_2 > p_1$), and only these 2 states are possible. Once the process is in state p_2, it stays there until corrected. During a workshift of a net length of N units of time—during which the process was engaged in production the entire time—there is a probability of p_N that the process will switch from state p_1 to state p_2.

Therefore, the probability that a jump will occur during a unit of time from state p_1 to state p_2, is p_{12}, such that $(1 - p_{12})^N = 1 - p_N$, providing

$$p_{12} = 1 - \sqrt[N]{(1 - p_N)} \tag{1}$$

It is assumed that whenever the change in the states in any time interval occurs, on the average it will happen at the mid-point of that time interval.

The process is controlled by a fraction-defective control chart which has its central line situated at the level of p_1, uses a sample of size n_i (where i is the sequential number of the sample), and has symmetrical control limits placed at K_i standard deviations from the central line. It is assumed that there is no further sampling, inspection or sorting of the manufactured goods.

When, and only when, a sample fraction-defective falls above the upper control limit (providing an indication of assignable causes) the manufacturing process is stopped, and a search is performed to locate the assignable cause(s) and to eliminate its (their) reappearance. If it turns out that it was just a false alarm, the process will be idle for a duration of t_f units of time, and the search will cost C_f. However, if the alarm was found to be a real one, it is assumed that the assignable causes are located and eliminated (culminating in a perfect reset), the process will be idle for a duration of t_r, and the search, repair and reset will cost C_r.

The average time the process is idle (due to false alarm or due to reset) following the ith sample, τ_i, is:

$\tau_i = t_f$ (Probability of false alarm on the ith sample)

$\quad + t_r$ (Probability of correct indication of out-of-control on the ith sample)

$$\tau_i = t_f P\left({}^ip > UCL_p|p_1\right) \cdot W_i + t_r P\left({}^ip > UCL_p|p_2\right) \cdot (1 - W_i) \quad , \tag{2}$$

where

$\quad {}^ip \quad = \quad$ Fraction of defectives of sample i ,

$\quad W_i \quad = \quad$ Probability that the production at the time of taking the ith sample is in state 1 (producing p_1 fraction defectives),

$\quad UCL_p \quad = \quad$ Upper control limit of a fraction-defective control chart

and

$$UCL_p = p_1 + K_i \sqrt{\frac{p_1(1 - p_1)}{n_i}} \tag{3}$$

Whereas the sample fraction-defective is distributed binomially,

$$P({}^ip > UCL_p|p_1) \quad = \sum_{x > n_i\left[p_1 + K_i\sqrt{\frac{p_1(1-p_1)}{n_i}}\right]}^{n_i} \binom{n_i}{x} p_1^x(1 - p_1)^{n_i - x} \quad , \text{ and} \tag{4}$$

$$P({}^ip > UCL_p|p_2) \quad = \sum_{x > n_i\left[p_1 + K_i\sqrt{\frac{p_1(1-p_1)}{n_i}}\right]}^{n_i} \binom{n_i}{x} p_2^x(1 - p_2)^{n_i - x} \tag{5}$$

The timing of the ith sample, measured from the start of the shift, is t_i, and it is illustrated in Figure 1.

172

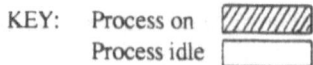

KEY: Process on
Process idle

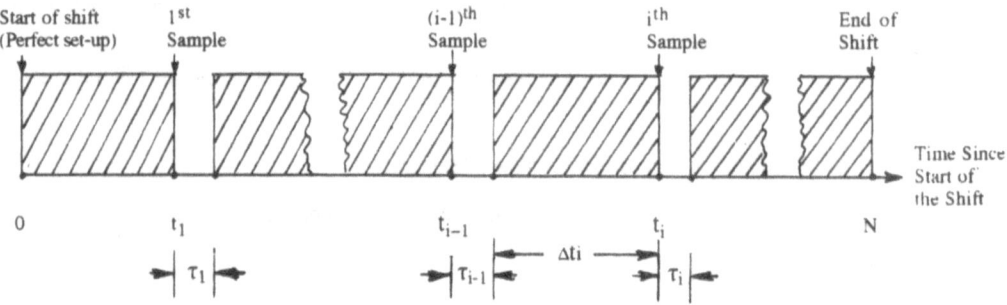

| Start of shift (Perfect set-up) | 1st Sample | $(i-1)^{th}$ Sample | i^{th} Sample | End of Shift |

FIGURE 1: The timing of the samples during the whole shift.

The probability that the production at the time of taking the ith sample is in state 1, W_i, is composed of 2 components:

(a) Probability that the $(i-1)^{th}$ sample is taken while the process is in state 1, and that during the production interval of Δt_i no jump from state 1 to state 2 has occurred, and where

$$\Delta t_i = t_i - t_{i-1} - \tau_{i-1} \tag{6}$$

and

(b) Probability that the $(i-1)^{th}$ sample is taken while the process is in state 2, and that the $(i-1)^{th}$ sample correctly indicates that the process is in state 2 (generating as assumed a perfect reset), and also that during the following production interval of Δt_i no jump from state 1 to state 2 occurred.

Thus,

$$W_i = W_{i-1}(1-p_{12})^{\Delta t_i} + (1-W_{i-1})P[^{(i-1)}p > UCL_p|p_2] \cdot (1-p_{12})^{\Delta t_i} \quad , \quad \text{and} \tag{7}$$

$$W_1 = (1-p_{12})^{t_1} \quad \text{since at start of shift the process is always in state 1.} \tag{8}$$

The probability that the process will be in state 1 (producing p_1 fraction-defectives), when the production resumes after the ith sample at time $t_i + \tau_i$, Q_i , is

$$Q_i = W_i + (1-W_i)P(^i p > UCL_p|p_2) \tag{9}$$

3. THE DYNAMIC PROGRAMMING MODEL

Assume that during a time interval of length T, which commences with the start of the shift, NO sample is taken. Then, the expected profit for the period of length T (with no sample), $\pi_0(T)$, is composed of the expected revenue generated when the process is in state 1, and the expected revenue generated when the process jumps to state 2, less the costs incurred. It should be remembered that if and when the process jumps from state 1 to state 2, it occurs at the middle of the time interval considered. Thus,

$$\pi_0(T) = BT\left\{(1-p_{12})^T\left[(1-p_1)R_c + p_1 R_{dc}\right] + \left[1-(1-p_{12})^T\right]\left[\frac{(1-p_1)+(1-p_2)}{2}R_c + \left(\frac{p_1+p_2}{2}\right)R_{dc}\right]\right\}$$

$$-\left\{C_2 + C_1 BT\right\} \qquad \text{for } T = 1, \dots, N. \quad , \qquad (10)$$

where

B $\quad$ = $\quad$ Number of units (conforming __and__ defectives) produced per unit of time,

R_c $\quad$ = $\quad$ Net revenue received from selling a conforming unit,

R_{dc} = $\quad$ Net revenue received from selling a defective unit as a conforming unit (it may be negative),

C_1 $\quad$ = $\quad$ The variable cost of manufacturing one unit (conforming or defective), including the cost of the raw material, and

C_2 $\quad$ = $\quad$ The fixed cost of manufacturing, per shift.

The terms in the first and second __heavy__ square brackets in equation (10) represent the generated revenue by the process, per unit, when the process is in state 1 and state 2, respectively.

Assume now, that during a time interval of T which commences with the start of the shift there is just one sample. That single sample is of size n_1, and is taken at time t_1 after the start of the shift. The expected profit for the whole period of length T with the one sample at time t_1, $\pi_1(T)$, can be expressed in a recursive fashion having 2 components:

(a) The profit from producing for a period of t_1 with __NO__ sample taken during t_1, i.e., $\pi_0(t_1)$, and

(b) The contribution to profit from producing for a period of $T - t_1 - \tau_1$ after taking the 1st sample, while taking into account the cost of the sampling.

Note that it is assumed that the defective items found in the sample are never remixed with the rest of the units, but rather they are sold as defective units generating a revenue of R_d per unit. Thus,

$$\pi_1(T) = \pi_0(t_1) \qquad (11)$$

$$+ Q_1\left\{(1-p_{12})^{T-t_1-\tau_1}\cdot\left[(1-p_1)R_c + p_1 R_{dc}\right]\right.$$

$$+\left[1-(1-p_{12})^{T-t_1-\tau_1}\right]\left[\left(1-\frac{p_1+p_2}{2}\right)R_c + \left(\frac{p_1+p_2}{2}\right)R_{dc}\right]\right\}B(T-t_1-\tau_1)$$

$$+(1-Q_1)\left\{(1-p_2)R_c + p_2 R_{dc}\right\}B(T-t_1-\tau_1)$$

$$+(R_d - R_{dc})\left\{n_1 p_1 W_1 + n_1 p_2(1-W_1)\right\}$$

$$-\left\{C_1 B(T-t_1-\tau_1) + C_4 + C_3 n_1\right\}$$

where

C_3 = The variable cost of sampling, per unit inspected, and

C_4 = The fixed cost of sampling, so that

$$C_4 = \begin{cases} S & \text{if} \quad n_i > 0 \\ 0 & \text{if} \quad n_i = 0 \end{cases} \tag{12}$$

The last ornamented bracket term in equation (11) contains the cost component of manufacturing during $T - t_1 - \tau_1$, and the cost of sampling at time t_1. The next to the last ornamented bracket term consists of the expected member of defective items found in the 1st sample, and it is multiplied by the revenue difference per unit (between defective unit sold as defective and defective unit sold as conforming). This is performed in order to adjust the revenue contained in $\pi_0(t_1)$ for the fact that each of the defectives found in the 1st sample (from the production at the end of time t_1) generates a revenue of R_d instead of R_{dc}.

Let us define the maximal value of $\pi_1(T)$ calculated over all possible combinations of values of the triplets t_1, n_1, and K_1, as $M\pi_1(T)$:

$$M\pi_1(T) = \underset{t_1, n_1, K_1}{\text{Max}} \pi_1(T) \qquad \text{for } T = 2, \ldots, N. \tag{13}$$

Hence, if during a time interval of T (which commences with the start of the shift) there are i samples, where the ith sample of size n_i is taken at time t_i after the start of the shift, and the upper control limit of the control chart is placed K_i standard deviations from p_1, then the Maximal expected profit for the whole period T will be $M\pi_i(T)$:

$$M\pi_i(T) = \underset{t_i, n_i, K_i}{\text{Max}} \Bigg\{ M\pi_{i-1}(t_i) + \tag{14}$$

$$+ Q_i \Bigg\{ (1 - p_{12})^{T - t_i - \tau_i} \cdot \left[(1 - p_1)R_c + p_1 R_{dc} \right]$$

$$+ \left[1 - (1 - p_{12})^{T - t_i - \tau_i} \right] \left[\left(1 - \frac{p_1 + p_2}{2} \right) R_c + \left(\frac{p_1 + p_2}{2} \right) R_{dc} \right] \Bigg\} B(T - t_i - \tau_i)$$

$$+ (1 - Q_i) \{ (1 - p_2)R_c + p_2 R_{dc} \} B(T - t_i - \tau_i)$$

$$+ (R_d - R_{dc}) \{ n_i p_1 W_i + n_i p_2 (1 - W_i) \}$$

$$- \left\{ C_1 B(T - t_i - \tau_i) + C_4 + C_3 n_i \right\} \Bigg\} \qquad \text{for } T = i + 1, \ldots, N .$$

Assuming that $M\pi_i (T = N)$ is unimodal in i (as has been the case with all cases solved), the optimal number of samples (and their characteristics) will be obtained at the highest possible stage i, for which

$$M\pi_{i-1}(T=N) < M\pi_i(T=N) \tag{15}$$

It will then provide the optimal number of samples, i_0, and the timing, size and distance of control limit for each one of them:

$$(t_1, n_1, K_1), (t_2, n_2, k_2), \text{.......} , (t_{i_0}, n_{i_0}, K_{i_0}).$$

4. SOLUTION

A. General Case

At each stage i, $i \geq 1$, for each value of T (where T ranges from $T = i + 1$ to N), $\pi_i(T)$ is calculated for each combination of the possible values of the <u>triplets</u> t_i, n_i and K_i. The obtained values of $\pi_i(T)$ are then compared with each other, and the triplet of t_i, n_i and K_i that provides $M\pi_i(T)$ (the maximal value of $\pi_i(T)$) for that specific T is determined. Thus, at stage i we have to save in the memory, for each value of T, a vector containing the $M\pi_i(T)$ and the corresponding values of the triplet. t_i, n_i and K_i.

Note, that if in order to reduce the number of comparisons we consider for K_i only the possible values of $K_i = 0.5, 1, 1.5 \ 2, 2.5, 3.$, while n_i may assume any discrete number between 3 and 400, and t_i may assume (at its maximum) any number between 2 and 480 (accuracy of one minute in an 8 hour work shift) — we still obtain about $6 \times 400 \times 480 = 1152 \times 10^3$ comparisons at the maximum. On the average, for each value of T we would obtain less than 576×10^3 comparisons.

Thus, for realistic general case solution which would overcome the "curse of dimensionality", the values of n_i should be considered only in steps of 10 or 20, an action which would reduce the average number of comparisons for each T to about 57×10^3 or 28×10^3, respectively. This amount of calculations is already tolerable, and the optimal values of the triplets t_i, n_i and K_i can be found for each of the stages $i = 1, 2, ... , i_0$. Further savings can be achieved if t_i is considered only in intervals of 5 or 10 minutes, reducing the average number of comparisons for each T to about $3,000$ for the latter case. This would generate — needing now the evaluation, on the average, for only 24 different values of T in each stage — of only about a total of $24 \times 3,000 = 72,000$ comparisons for each stage. This amount of computations is handleable by a personal computer.

However, it is impracticable to have at each sample a control-limit on the control-chart which is away from the central line a different multiple of the standard deviation. Moreover, the industry has good experience with the traditional deeply-rooted common practice of using control limits 6 standard deviations apart. Hence, the adoption of the practical case of $K_i = 3$ for all i will reduce the number of comparisons (for each T) to one-sixth of its previous size.

B. Special case

Differing sample sizes pose additional problems. They require the use of a fraction-defective control-cart with unequal control-limits, or the use of a "stabilized" control-chart where the vertical axis is expressed in standard deviation units. Thus, it seems practical to consider the policy where all the samples have equal sizes, and only their spacing is different.

In this latter practical situation (coupled with equal values of $K_i = 3$ for all i), the solution of the Dynamic Programming model becomes slightly different. At each stage i, $i > 1$, for each value of

T . $\pi_i(T)$ is calculated for each combination of the possible values of the <u>pairs</u> t_i and n . For a given value of T , and for a given value of n , the obtained values of $\pi_i(T)$ are compared with each other. and that value of t_i is determined which generates the maximal value of $\pi_i(T)$. Let us denote the maximal value of $\pi_i(T)$ for the specific value of T and for the specific value of n as $M\pi_i(T, n)$. As a result, equation (16) that follows replaces (14), and equation (17) replaces (15).

$$
\begin{aligned}
M\pi_i(T,n) = \underset{t_i}{Max}\Bigg\{ &M\pi_{i-1}(t_i,n) \\
&+Q_i\Bigg\{ (1-p_{12})^{T-t_i-\tau_i}\cdot\left[(1-p_1)R_c + p_1 R_{dc}\right] \\
&\quad +\left[1-(1-p_{12})^{T-t_i-\tau_i}\right]\left[\left(1-\frac{p_1+p_2}{2}\right)R_c +\left(\frac{p_1+p_2}{2}\right)R_{dc}\right]\Bigg\}B\left(T-t_i-\tau_i\right) \\
&+(1-Q_i)\left\{(1-p_2)R_c + p_2 R_{dc}\right\}B\left(T-t_i-\tau_i\right) \\
&+(R_d - R_{dc})\left\{np_1 W_i + np_2(1-W_i)\right\} \\
&-\left\{C_1 B\left(T-t_i-\tau_i\right)+C_4+C_3 n\right\}\Bigg\} \qquad \text{for} \quad T=i+1,...,N .
\end{aligned}
$$
(16)

$$
\underset{n}{Max}\left\{M\pi_{i-1}(T=N,n)\right\} < \underset{n}{Max}\left\{M\pi_i(T=N,n)\right\} .
$$
(17)

The solution of equations (10), (16) and (17) provides the optimal values of n , and of t_i for $i=1$, ... , i_0 .

Thus, at stage i we have to save in the memory a $T \times n$ matrix which contains (for each combination of the values T and n) a vector of $M\pi_i(T, n)$ and the corresponding value of t_i .

Hence, at each stage, for each entry in the matrix, we will have to make, on the average, 240 comparisons in the "unlimited search" case. If the sample size is considered only in steps of 10 or 20 , the size of the matrix, on the average, will be about 40×240 or 20×240 , respectively. Thus, the total number of comparisons at each stage will remain the same, except for the fact that the end of each stage requires about 40 or 20 additional comparisons due to calculations of equation (17) . (In equation (17) the maximum profits for different values of equal n in all stages are compared.)

As in the general case, consideration of t_i only in steps of 10 will further reduce the number of comparisons to one tenth, requiring for each stage, on the average, (for n in steps of 20 , and t_i in steps of 10) of only 5,780 comparisons.

5. NUMERICAL EXAMPLE

The equal–sample-size case was evaluated for the traditional situation of $K_i = 3$ for all i, and for the economic and probabilistic data conditions as follows:

$p_i = 0.01$	$p_2 = 0.03$	$p_N = 0.4$ per work shift
$t_r = 2$ min.	$t_f = 1$ min.	
$C_r = 1$ \$	$C_f = 2$ \$	

$C_1 = 1$ \$/unit $\quad C_2 = 10$ \$/shift $\quad C_3 = 2.5$ \$/unit $\quad C_4 = 20$ \$/sample

$B = 2$ units/minute

$R_d = 3$ \$/unit $\quad R_{dc} = -200$ \$/unit $\quad R_c = 4$ \$/unit

Using equations (10), and (16), we obtained:

$$
\begin{aligned}
\pi_0(T = 480) &= 128.24 \text{ \$/shift} \\
\pi_1(T = 480) &= 625.32 \\
\pi_2(T = 480) &= 706.79 \quad \text{<--- Maximum} \\
\pi_3(T = 480) &= 682.99 \\
\pi_4(T = 480) &= 658.76 \\
\pi_5(T = 480) &= 634.46 \\
\pi_6(T = 480) &= 609.69
\end{aligned}
$$

Thus, the results validate the unimodality assumption, as we have observed in all cases investigated.

The optimal solution calls for the first sample at $t_1 = 361$ min. of $n_1 = 400$ units, and the second sample at $t_2 = 456$ min. of $n_2 = 400$ units, and it provides a profit of 706.79 \$/shift. Note, that in the present case with 3 or more samples per shift, the best could be obtained with sample sizes of 20, but the resulting profit would be less than the optimal. On the other hand, with less than 2 samples, the best obtainable profit rapidly and significantly falls short of the maximum.

Using a different set of data, the optimum solution called for no sample, i.e., π_0 (T=480) was maximal, and π_i(T=480) monotonically decreased with increase in i.

The above results indicate that each numerical case should be evaluated individually using the Dynamic Programming formulation, and that analysis is not providing any advance hint about the location of the optimal solution, or the magnitude of savings in using optimal versus non-optimal policies.

It might be interesting to see the gain of the optimal policy over an equal-spaced equal–sample-size policy. The highest profit generated from this latter "constant design" strategy can be obtained from the solution of equations (10), (14), and (15) (or equations (10), (16), and (17)), where we impose the conditions that

$$
\begin{aligned}
n_i &= \text{constant for all } i \text{ , and} \\
t_i &= \left(\frac{i}{i+1}\right) T \quad .
\end{aligned}
$$

REFERENCES

[1] BAKER, K.R. (1971): Two process models in the economic design of an $\bar{X}$ chart, AIIE Transactions, Vol. 3, No. 4.

[2] BARLOW, R.E., HUNTER, L.C. and PROSCHAN, F. (1963): Optimum checking procedures, J. Society of Industrial & Applied Mathematics, Vol. 11, 1078-1095.

[3] BARLOW, R.E. and PROSCHAN, F. (1965): Mathematical theory of reliability. Pp. 107-118. Wiley: New York.

[4] CHIU, W.K. (1975): Economic design of attribute control charts, Technometrics, Vol. 17, 81-87.

[5] CHIU, W.K. and WETHERILL, B. (1974): A simplified scheme for the economic design of $\bar{X}$-Charts, Journal of Quality Technology, April, 63-69.

[6] DUNCAN, A.J. (1956): The economic design of $\bar{X}$-charts used to maintain current control of process, Journal of American Statistical Association, Vol. 51, 228-242.

[7] GIBRA, I.H. (1971): Economically optimal determination of the parameters of an $\bar{X}$ chart, AIIE Transactions, Vol. 3, No. 4.

[8] GOEL, A.L., JAIN, S.D., and WU, S.M. (1968): An algorithm for the determination of the economic design of $\bar{X}$-charts based on Duncan's Model, Journal of the American Statistical Association, Vol. 63.

[9] HALL, R.I., and EILON, S. (1963): Controlling production processes which are subject to linear trends, Operational Research Quarterly, Vol. 14, No. 3, 279-289.

[10] KELLER, J.B. (1974): Optimum checking schedules for systems subject to random failure, Management Science, Vol. 21, No. 3, 256-266.

[11] KNAPPENBERGER, H.A. and GRANDAGE, A.H. (1969): Minimum cost quality control tests, AIIE Transactions, Vol. 1, No. 1.

[12] LADANY, S.P. (1973): Optimal use of control charts for controlling current production, Management Sciences, Vol. 19, No. 7, 763-772.

[13] LADANY, S.P. and BEDI, DINA N. (1976): Selection of the optimal setup policy, Naval Research Logistics Quarterly, Vol. 23, No. 2, 219-233.

[14] LUSS, H. (1983): An inspection policy model for production facilities, Management Science, Vol. 29, No. 9, 1102-1109.

[15] LUSS, H. and KANDER, Z. (1974): Inspection policies when duration of checkings is non-negligible, Operational Research Quarterly, Vol. 25, No. 2, 299-309.

[16] MONTGOMERY, D.C. (1980): The economic design of control charts: A review and literature survey, Journal of Quality Technology, Vol. 12, 75-87.

[17] MONTGOMERY, D.C. and KLATT, P.J. (1972): Economic design of T^2 control charts to maintain current control of a process, Management Science, Vol. 19, No. 1.

[18] MUMFORD, A.G. (1981): Comparison among certain inspection policies, Management Science, Vol. 27, No. 3, 260-267.

[19] RAZ, Z. and LADANY, S.P. (1991) Optimal inspection policy for imperfect inspection in discrete times, <u>Production Planning and Control</u>, (forthcoming).

[20] SACKROWITZ, H. and SAMUEL-CAHN, E. (1974): Inspection procedures for Markov chains, <u>Management Science</u>, Vol. 21, No. 3, 261-270.

[21] SANIGA, E. and MONTGOMERY, D.C. (1981): Economical quality control policies for a single cause system, <u>AIIE Transactions</u>, Vol. 13, No. 3, 258-264.

[22] SENGUPTA, B. (1979): Inspection procedures when failure symptoms are delayed, <u>Operations Research</u>, Vol. 27, No. 4, 768-776.

[23] WATTANAPANOM, N. and SHAW, L. (1979): Optimal inspection schedules for failure detection in a model where tests hasten failures, <u>Operations Research</u>, Vol. 27, No. 2, 303-317.

Importance of Process and Control
in Statistical Process Control

A. K. Shahani, Southampton, England

1. INTRODUCTION

We consider a process, or a system, working in time that can be in a number of states. Some states are more desirable than others. The process has to be inspected to detect undesirable states and to avoid the penalties of operating in these undesirable states.

In the manufacturing industry control of processes usually includes the inspection of the product produced by the processes. Indeed most of the work on Statistical Process Control (SPC), starting from the pioneering work of W.A. Shewhart, is in terms of sampling units of products produced by a process.

Shewhart's work continues to be a common basis for SPC in practice. The methodology here is the statistical testing of an hypothesis. With $\theta = \theta_0$ as the target, SPC is often viewed as a repeated testing of the hypothesis $H_0: \theta = \theta_0$ at each inspection. Thus the inspection scheme essentially requires the definition of the sample size, test statistic, and the critical region. A graphical implementation of this procedure in the form of control charts is simple and helpful. Simplicity and wide ranging use are the strengths of this methodology.

A major weakness of this methodology is that it does not take into account the fundamental point that the process is operating in time.

This paper argues that in designing process inspection schemes more attention should be given to:

(1) The dynamics of the process

(2) An explicit consideration of time

Some theory of SPC with process dynamics is given and a practical case study is described. In the absence of a model for process dynamics, time can be considered, explicitly and easily, through a conditional analysis and this approach is illustrated by an example. Control can be achieved through a variety of techniques, however the role of people is fundamental.

2. A THEORY OF SPC WITH PROCESS DYNAMICS

With θ as the process quality, a necessary condition for the existence of a process inspection and control problem is that θ can change over time. Further, some values of θ are preferable to others. If θ does not change we do not need SPC. If the process quality θ is a constant but is not satisfactory, we have a variety of problems and examples are: possible screening of useable product, the design of experiments for improving θ, and major investment in new machines.

Frontiers in Statistical Quality Control 4
Ed. by Lenz et al.
© Physica-Verlag Heidelberg 1992

In some cases the dynamics of the process are well understood and mathematical models for the process lead to a deterministic model for θ over time, say $\theta(\tau)$. Usually in such cases we have a pure process control problem, without any major inspection efforts, and the solution is the definition of an appropriate action to be taken at time τ, say $a(\tau)$. For example $a(\tau)$ could be needed to keep $\theta(\tau)$ within some defined limits. As an illustration suppose $\theta = \theta_0$ at $\tau = 0$ and a model for θ is:

$\theta(\tau) \geq \theta_0$, $\qquad \tau \geq 0$

$\theta(0) = \theta_0$

$\theta(\tau) = \theta_0 + b\tau$, $\quad b > 0$.

Further suppose that acceptable range of θ is $\theta_0 \leq \theta(\tau) \leq \theta_1$. In this case, one control solution would be, action at time $\tau = \dfrac{\theta_1 - \theta_0}{b}$, and the action could be to reset θ to θ_0, and τ to 0.

The need for Statistical Process Control arises when there is some uncertainty about $\theta(\tau)$. Typically this uncertainty is the result of an insufficient understanding of the dynamics of the process and imprecise inspection. For example, we may not know how θ changes with time; further, inspection results are typically from a small sample from a distribution with θ as parameter(s) so that θ cannot be measured precisely.

There are two approaches for dealing with uncertainty about $\theta(\tau)$.

One idea is to use an appropriate model for $\theta(\tau)$. Here some understanding of the dynamics of the process is clearly essential. In some cases $\theta(\tau)$ could be modelled as an appropriate stochastic process. Clearly if a particular model for $\theta(\tau)$ is appropriate then that model should be used in designing SPC schemes. In the absence of a special model, it is suggested that the following "shock model" should be considered.

At time 0, the process starts in the desired state θ_0. Subsequently, inspections have to be scheduled at times u_i, $i = 1, 2, 3 \ldots$ for obtaining information about $\theta(u_i)$. The process will suffer from a shock at time t which is a realisation of a random variable T. In the basic version of the shock model, the process has two states and at time t, θ_0 changes to θ_1. More generally θ is a known deterministic function of time after the instant of the shock at time t. Now, the earliest time at which the change from θ_0 can be detected is at scheduled time $u_i \geq t$ so that the process will operate in an undesirable state θ for at least $(u_i - t)$. One of the considerations in the calculation of u_i would be to achieve a good balance between the amount of inspection and the delay in the detection of a change from θ_0. The case of inspections yielding precise information about θ has been analysed using simple linear cost functions and a necessary condition for calculating optimal u_i has been obtained (Barlow et al 1963). These optimal u_i exist for a particular class of probability density functions of T and this class includes T with non-decreasing failure rate functions. When the failure rate function is increasing the optimal u_i are such that $d_i = u_i - u_{i-1}$ is a decreasing sequence. In practice a schedule of u_i with decreasing d_i is often unacceptable. This observation indicates that the simple linear cost functions are somewhat inadequate for they do not take the concept of practically acceptable inspection schedules into account.

This shock model has been considered by a number of workers at the University of Southampton (Munford and Shahani 1972, 1973; Shahani 1981; V. de Senna & Shahani 1986; Shahani et al 1987) and the conclusion is that a very good practical schedule of inspection times is obtained from

$$u_i = m + (i - 1) \, c, \qquad i = 1, 2, 3 \dots . \quad u_0 = 0. \tag{1}$$

Note that $c = m$ is the case of periodic inspections $u_i = i \, c$, $i = 1, 2, 3 \dots$. Generally if T has an increasing failure rate then $m > c$, and $m < c$ for the case of a decreasing failure rate. If T is negative exponential, that is if T has a constant failure rate, then $m = c$.

This shock model can be used with a variety of rules for stopping inspection and adjusting the process. In all the cases the time origin is the instant of adjusting the process to θ_0 and u_i are measured from this time origin. Inspection may, or may not, measure θ precisely. The simplicity of the Shewhart control chart rules continues to be attractive in practice and with this rule the inspection scheme corresponds to a series of independent trials, carried out at time u_i, each of which can have one of two outcomes, say S_0, S_1. The occurrence of S_1 is taken to be evidence for $\theta \neq \theta_0$. Note that the introduction of outcomes S_0, S_1 implies that inspection is imprecise and we define the following error probabilities

$$P \text{ (outcome } S_1 | \theta_0) = a \tag{2}$$

$$P \text{ (outcome } S_0 | \theta \neq \theta_0) = b(\theta) \tag{3}$$

The probabilities a and b could be controlled by the choice of test statistics, sample size, and the definitions of S_0, S_1. We suppose that the cost of corrective action is sufficiently large to warrant the use of a precise inspection for verifying that $\theta \neq \theta_0$ as soon as S_1 occurs. The reason for not using this precise inspection all the time is, of course, the high cost of this precise inspection. We suppose that the verification tests are carried out swiftly and that these tests cause no delays to any further necessary scheduled tests.

2.1 A BASIC ANALYSIS

Appropriate mathematical results, and an application, for the basic two state model have, already been reported (Shahani et al 1987) and this case corresponds to a constant b. We now consider the case of the shock model where θ changes in a known, deterministic, manner from θ_0 after receiving a shock. Let $\theta(\tau)$ be the model for changes in θ. Thus $\theta(\tau) = \theta_0$ for $0 < \tau < t$, where t is the realisation of a random variable T, and $\theta(\tau)$ is a known function for $\tau > t$.

Now the probability of missing the change from θ_0 at scheduled inspection $u_i > t$ is

$$P(\text{Outcome } S_0 \text{ at time } u_i | \theta \neq \theta_0) = b(\theta(u_i)) \tag{4}$$

and we use the notation $b_i = b(\theta(u_i))$ for convenience.

Let $f(t)$ be the probability density function of T and let

$$F(t) = \int_0^t f(x) \, dx, \qquad S(t) = 1 - F(t). \tag{5}$$

The probability of a false alarm has been defined by equation (2). Each false alarm will result in a precise verification test for checking whether θ has changed from θ_0 and this is an expensive test. The number of false alarms is therefore an important random variable.

Let H be the number of false alarms.

$$\text{Now} \quad P(H = h \,|\, u_{i-1} < T \le u_i) = \binom{i-1}{h} a^h (1-a)^{i-1-h} \tag{6}$$

so that $E(H) \,|\, u_{i-1} < T \le u_i) = (i-1) a$

$$\text{Further} \quad P(H = h) = \sum_{i=1}^{\infty} \binom{i-1}{h} a^h (1-a)^{i-1-h} [F(u_i) - F(u_{i-1})]$$

$$\text{and} \quad E(H) = \sum_{i=1}^{\infty} (i-1)a[F(u_i) - F(u_{i-1})]$$

$$= a \sum_{i=1}^{\infty} S(u_i) \tag{7}$$

Let N be the number of scheduled tests between successive adjustments of the process. N is a measure of the amount of inspection and it is clearly an important random variable.

Let $A_{i,n}$ be the event that $u_{i-1} < T \le u_i$ and $N = n$

$$\text{Now} \quad P(A_{i,n}) = [F(u_i) - F(u_{i-1})] \, b_i b_{i+1} b_{i+2} \ldots b_{n-1} (1 - b_n)$$

$$\text{So that} \quad P(N = n) = \sum_{i=1}^{n} P(A_{i,n}) \tag{8}$$

The delay in the detection of a change from θ_0 is $U_N - T$ and the mean value of this delay is $E(U_N) - E(T)$.

$$\text{Now} \quad E(U_N) = \sum_{n-1}^{\infty} u_n P(N = n) \tag{9}$$

$$\text{Also} \quad E(N) = \sum_{n-1}^{\infty} n P(N = n) \tag{10}$$

2.2 USE OF THE ANALYSIS

We discuss the use of this analysis through a particular example. In this discussion we suppose that $f(t)$ is known. Of course in practice $f(t)$ is not known and we will deal with practical difficulty in Section 3 through a case study.

We suppose that the idea of an optimal process inspection scheme which is obtained by minimising an appropriate cost function is not a practical one. The problem now is the choice of the many variables that are involved in the definition of a process inspection scheme.

Suppose $\theta \geq \theta_0$ is the mean value of a normal variate with a known variance σ^2. A sample of size k will be taken at the scheduled inspection times u_i and the sample mean $\bar{x} = \sum_{j=1}^{k} \frac{x_j}{k}$ will be used to define the outcomes S_0, S_1 as follows

S_0 occurs if $\bar{x} < \theta_0 + L \ \sigma/\sqrt{k}$, -- S_1 occurs if $\bar{x} \geq \theta_0 + L \ \sigma/\sqrt{k}$

Consider the value of b at a defined $\theta > \theta_0$ say $\theta = \theta_1$. The four variables a, $b(\theta_1)$, L, k are such that if two of them are specified, the other two can be computed. Now the false alarm probability a controls the important variable H and a knowledge of the cost due to false alarms may provide a guide to a good range of the values of this probability and further there may be practical reasons for limiting k to a small set of values and in this case the search for a good process inspection scheme would start with a few combinations of (a, k) values. If there are no practical considerations for guiding the choice of k, the search for a good process inspection scheme would start with a few combinations of (a, $b(\theta_1)$) values.

We have suggested the use of $u_i = m + (i - 1) \ c$ and this requires the choice of the variables c, m. If a, $b(\theta)$ are fixed, c and m will influence N and U_N - T. For example, small values of c, m will mean a large value of E(N) and a small value of $E(U_N$ - T). A good way of choosing c, m would be to compute E(N) and $E(U_N$ - T) for chosen combinations of (c, m). It will be seen that values of (c, m) generate points in the E(N), $(E(U_N$ - T) plane and that these points fall in a region bounded by a curve that exhibits the law of diminishing returns. Figure 1 illustrates the consequences of the choice of particular values of (c, m); the values of $E(U_N$ - T) and E(N) that result from the particular choice of (c, m) would yield the position of the plotted points. This figure is not scaled and this emphasises the general nature of the consequences of the choice of (c, m).

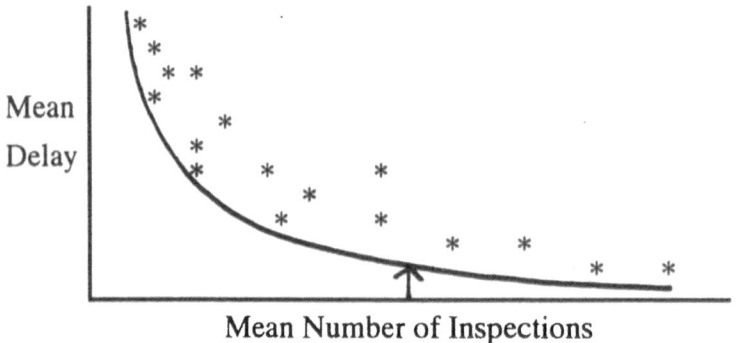

Figure 1 Mean delay vs mean number of inspections

The law of diminishing returns is clearly seen for large increases in $E(N)$ are needed to reduce $E(U_N - T)$ below the level indicated by the arrow.

At first sight the amount of information needed and the number of variables involved seem to make the choice of a good process inspection rather daunting. However, a well designed microcomputer program with good graphical screens will rapidly evolve the necessary process inspection schemes.

3. A CASE STUDY

This case study has been chosen to illustrate that good process inspection schemes using the shock model theory can be obtained for real practical problems. The case study has been chosen for a discussion about insufficient knowledge of T, which is the time for the transition from $\theta = \theta_0$ to $\theta \neq \theta_0$.

The problem was posed by Exxon Chemical Ltd and it was concerned with the scheduling of inspections for the detection of excessive corrosion in a very large network of metal pipes. These pipes exist in a variety of environments, for example hot and humid, exposed to weather, and buried underground. The pipes are protected by lagging and aluminium cases, however this protection does not stop the pipes corroding. Too much corrosion will, if left undetected, eventually result in holes in the pipes. The pipes carry a wide range of liquid products and the consequences of a leak due to a hole in a pipe are very serious indeed. Scheduled inspections for the detection of corrosion are expensive and very frequent inspections must be avoided.

The thickness of a pipe is specified by the designer and it takes the use of the pipe and its operating environment into account. The process quality θ is the thickness and generally corrosion decreases θ over time. The corrosion process is not understood sufficiently for the use of a predictive model for estimating θ at any time from the time origin of the instant of installing a new pipe. Discussions with the engineer in charge resulted in the understanding that:

(1) Reduction in the original pipe thickness of 50%, or more, should be regarded as a serious change in pipe quality

(2) The rate at which corrosion takes place is very variable and unpredictable. Environmental conditions influence corrosion. Data for classifying pipes according to environmental conditions is available.

(3) The idea of using a two state model with the definitions θ_0: Good state of pipe, θ_1: Bad state of pipe and T as the transition between θ_0 and θ_1 is a good practical one. θ_1 would be the state when the original pipe thickness is reduced by at least 50%.

(4) Pipe thickness can be measured very precisely at each of the scheduled inspection. Thus the process quality θ is known without error.

(5) Because of the large number of pipes and the difficulties involved in inspections, many people, and contractors, are involved in the inspection process. Inspections have to be scheduled and the schedules have to be arranged, and made known, some years before an inspection. Thus the idea of a dynamic inspection schedule in which different pipes

are inspected at different times depending on the thickness measured at the current inspection is not a practical idea.

(6) Use of $u_i = m + (i - 1) c$, which includes the case of $u_i = ic$, is a good idea.

Choice of a model for T

In this problem we have the common difficulty that $f(t)$ cannot be derived from some mathematical model of corrosion. Further there is insufficient data for deducing $f(t)$ from data alone. We remark here that there are dangers in deducing $f(t)$ from data alone.

Corrosion is, in general, a cumulative process and a reasonable inference is that the conditional probability of the transition from θ_0 to θ_1 in the next time period, given θ_0 at the current time, is at least non-decreasing. This probability is likely to be an increasing function. Given these observations the choice of Weibull T (Weibull 1951) seems to be appropriate. That is, we choose

$$f(t) = \frac{\beta}{\alpha} \left[\frac{t}{\alpha}\right]^{\beta-1} \exp -\left[\frac{t}{\alpha}\right]^{\beta} \tag{11}$$

The parameters α, β would depend on the environment of a particular pipe, and we expect $\beta \geq 1$. Available data for estimating α, β corresponded to observations on $(t, S(t))$ so that two pairs say $(t_1, S(t_1))$ and $(t_2, S(t_2))$ would be needed to estimate the two parameters α, β from the solution of the equations $S(t_1) = v$, $S(t_2) = w$. A detailed account of the choice of $S(t_1)$, $S(t_2)$ in inappropriate here and the interested reader may wish to start with a paper by Dubey (Dubey 1967).

Table 1 gives a selection of the estimates obtained from the data

Estimates	Environment	
	Type I	Type II
α	58	29
β	1.3	1.4

Table 1 Estimated values of the Weibull parameters

Cost parameters were introduced even though the penalty cost of using a pipe in θ_1 is not easy to define. These costs served two purposes

(1) A range of cost values can be used to give some guidance for a rational choice of the budget for an inspection programme.

(2) The costs define a dummy variable which can be used to identify good combinations of c,m.

The cost parameters are that each inspection costs C_1 and the penalty of using a pipe in θ_1 is C_2 per unit time. Thus if the transition to θ_1 occurs at time t and θ_1 is detected at time $u_i \geq t$, the total cost, C, is

$$C = iC_1 + C_2 (u_i - t) \tag{12}$$

This is the linear cost function proposed by Barlow et al (Barlow et al 1963).

Now $E(C) = \sum_{i=1}^{\infty} \int_{u_{i-1}}^{u_i} [iC_1 + C_2(u_i - t)] \, f(t) \, dt$ (13)

and we seek m,c that minimise E(C).

Detailed analysis, including the role of $\dfrac{C_1}{C_2}$ as a dummy variable is available (Shahani 1991).

Discussions with the Engineer in Charge resulted in the development of a helpful microcomputer program for evolving, and exploring, inspection schedules. Details are available (Hanlon 1985, Shahani 1991). Here we mention that a user can input values of $[t_1, S(t_1)]$, $[t_2, S(t_2)]$ to estimate α,β and use these estimates in generating inspection schedules; or the user can input values of α,β and this option is helpful for exploring the sensitivity of the inspection schedule to α,β. Other options are computing c,m through using $\dfrac{C_1}{C_2}$ or using any desired c,m. A variety of numerical and graphical output is computed and examples are

"Optimal" c,m

Plot of $E(U_N - T)$ vs $E(N)$

Cost of inspection

Cost ratio of two inspection schedules

Illustrative inspection schedules for two of the many environments are given below. The time is measured in years

$u_i = 16 + 7(i-1)$, i = 1, 2, 3, ... Type I environment

$u_i = 13 + 5(i-1)$, i = 1, 2, 3, ... Type II environment

The Engineer in Charge was able to use this program to make decisions about good inspection schedules. Further he was able to provide a rational argument for budgeting expenditure on scheduled inspection to the management of Exxon Chemical Ltd.

4. A SIMPLE CONDITIONAL ANALYSIS FOR TAKING TIME INTO ACCOUNT

In the absence of any model for changes in θ, a simple way to take time into account is to define $T(\theta)$ as the conditional time between successive corrective actions for a given θ.

Let u_i, i = 1, 2, 3 ... be the scheduled inspections with $u_0 = 0$ as the instant of adjusting the process. Let $u_n(\theta)$ be time, for a given θ, at which the current series of inspections will stop and corrective action will be taken. In this particular case $T(\theta)$ has taken the value $u_n(\theta)$

With $u_i = m + (i - 1)c$, we have

$E(U_I) = m + c[E(I) - 1] = E[T(\theta)]$ (14)

The case of m = c gives

$E(U_I) = c \, E(I) = E[T(\theta)]$ (15)

E(I) will be recognised as the well known Average Run Length (ARL) (Aroian and Levene 1950)

of a process inspection scheme. Thus choice of a sampling frequency defined by c, or two sample frequencies defined by m and c, linked with the ARL function of a process inspection scheme is an easy way for an explicit consideration of time in the choice of a process inspection scheme. With this approach the dependent nature of the various variables that define a process inspection scheme is taken into account. The classical conditional probability approach ignores sampling frequency and it allows an independent choice of variables that really ought to be considered jointly. An example will illustrate these points.

Example

Let θ be the mean value of a normal variate with a known variance σ^2, and $\theta \leq \theta_0$. The target value is $\theta_0 = 1000$ and $\theta = 970$ is a particularly serious change. A simple process inspection is needed and the specification for $T(\theta)$ is

$$E[T(1000)] = 500 \text{ hours} \tag{16}$$

$$E(T(970)] = 10 \text{ hours} \tag{17}$$

We consider the median, x_k, of a sample of size $2k-1$ taken at times $u_i = ic$. The stopping rule is that as soon as any $x_k \leq L$, the process is stopped and adjusted.

Let $P(\theta) = P(x_k \leq L | \theta)$

Now $E(I) = \dfrac{1}{P(\theta)}$, and $E[T(\theta)] = \dfrac{c}{P(\theta)}$

Suppose $c = 0.5, 2.5, 5.0$ are under consideration. The specification on $E[T(\theta)]$ can now be translated into specification on E(I) and this is shown in Table 2.

$E[T(\theta)]$	$c = 0.5$	$c = 2.5$	$c = 5.0$
500	1000	200	100
10	20	4	2

Table 2   Values of E(I)

Published Tables of E(I) (Shahani 1971) yield the solutions for 2k - 1, L as shown in Table 3.

c	2k - 1	L
0.5	5	$1000 - 1.669\sigma$
2.5	9	$1000 - 1.054\sigma$
5.0	15	$1000 - 0.734\sigma$

Table 3   Definition of process inspection schemes

Table 3 demonstrates that sampling frequency and sample size ought to be considered jointly. Intuitively a smaller sample size means a greater sampling frequency; however, the relationship between sample size and sampling frequency is not obvious.

5. CONTROL: THE IMPORTANCE OF PEOPLE

A process inspection scheme monitors the quality and gives a signal for corrective action. Control of quality needs conscious actions. In this respect, SPC is not like Control Engineering solutions in which control is typically achieved automatically.

The Japanese emphasis on quality and the importance or trusing people, providing motivation, and making people responsible, is now increasingly regarded as being the right approach in the western world. A number of countries have tried Quality Circles; Total Quality Management is much talked about.

Good process control and process improvement will result from creating the conditions that encourage necessary corrective actions by the people who are involved with the process. Some companies in England are trying quality Circle type idea through Process Control and Improvement Teams. This team approach is an excellent organisation for control, and the necessary tools that such a team might need are many and varied. We mention Pareto analysis, Fishbone diagrams, Brain Storming, multi-factor experiments (on-line, and off-line). For a number of processes, the information provided by the usual control charts has to be supplemented by estimates of current quality level θ. Here some of the well established short term forecasting techniques would be helpful and we mention simple exponential smoothing with tracking signal (Trigg and Leach 1967, Batty 1969) and the Holt-Winters procedures (Chatfield 1978).

Thus one of the questions to be answered is, "for good control do we need to supplement a control chart which signals that $\theta \neq \theta_0$, with an estimate of the current θ?". If the answer is yes, appropriate software on a microcomputer is likely to be a good method of providing the necessary information.

Departmental loyalties and pressures often create difficulties in control at the level of the whole, or substantial part, organisation. As an illustration, we mention the problem of the quality of the effluent discharged in a river by a chemical plant. The plant has many departments and many processes. The company monitors the effluent quality by sampling at the discharge point. Poor quality samples result in Effluent Reports and these reports are meant to lead to actions for maintaining effluent quality. In a pilot study, our investigations indicated that the sampling procedure was not a good one from the point of view of control. The discharge point was too remote from the processes that could be the cause of poor quality effluent. But the fundamental difficulty was identified to be the lack of communication between the various departments. Our recommendation was to evolve an appropriate team of people. Duties of this team would include: improvements of sampling procedures, the acquisition of any necessary tools for data analysis; identification of appropriate chemical and other actions for improved control over effluent quality.

6. CONCLUDING REMARK

More attention to the important bits, namely Process and Control is needed in successful use of SPC.

7. ACKNOWLEDGEMENT

It is a pleasure to record my thanks to Dr. S. Winnik for his practical guidance during the corrosion study reported in Section 3.

8. REFERENCES

[1]   AROIAN, L.A. & LEVENE, H.·(1950): The effectiveness of quality control charts.
      J. Amer. Statist. Assoc., Vol. 45, 520-29.

[2]   BARLOW, R.E. et al. (1963): Optimum checking procedures.
      J. Soc. Ind. & Appl. Math., Vol. 11, 1078-1095.

[3]   BATTY, M. (1969): Monitoring an exponential smoothing forecasting system.
      Op. Res. Quart., Vol. 20, 319-326.

[4]   CHATFIELD, C. (1978): The Holt-Winters forecasting procedure.
      Applied Statistics,   Vol. 27, 264-278.

[5]   de SENNA, v., & SHAHANI, A.K. (1986): A simple inspection policy for the detection of
      failure.
      Euro. J. Op. Res. Vol. 23, 222-227.

[6]   DUBEY, S. (1967): Some percentile estimators for Weibull parameters.
      Technometrics, Vol. 9, 119-129.

[7]   HANLON, M. (1985): Inspection of a chemical plant.
      M.Sc. Dissertation, Mathematics Department, University of Southampton.

[8]   MUNFORD, A.G. & SHAHANI, A.K. (1972):   A nearly optimal inspection policy.
      Op. Res. Quart., Vol. 23, 373-379.

[9]   MUNFORD, A.G. & SHAHANI, A.K. (1973): An inspection policy for the Weibull case.
      Op. Res. Quart., Vol. 453-458.

[10] SHAHANI, A.K. (1971): A control chart based on sample median.
      The Quality Engineer, Vol. 35, 7-9.

[11] SHAHANI, A.K. (1981): choice of process inspection scheme: some basic considerations.
      Frontiers in Statistical Quality Control, Physica-Verlag, Wuerzburg-Vienna.

[12] SHAHANI, A.K. et al (1987). Two-test scheme for the detection of failure.
      Frontiers in Statistical Quality Control, 3. Physica-Verlag, Heidelberg.

[13] SHAHANI, A.K. (1991). Scheduling of inspections for monitoring the quality of systems.
      Preprint number 31, Mathematics Department, University of Southampton.

[14] TRIGG, D.W. & LEACH, A.G. (1967): Exponential smoothing with an adaptive response rate.
      Op. Res. Quart., Vol. 18, 53-60.

[15] WEIBULL, W. (1951): A statistical distribution of wide applicability.
      J. Appl. Mech. Vol. 18, 293-297.

Robustness and Flexibility of Constrained Economic Designs

E. M. Saniga, Newark, Delaware, USA

1. Introduction

Control charts are perhaps the most important methods available for the improvement and maintenance of product or service quality. The selection of sample size, control limit width and sampling frequency is called control chart design. The problem of control chart design has received much attention because design has behavioral, economic and quality implications. Consequently, there have been several general methodologies developed to improve upon the design suggested by Shewhart (1931), who originated the idea of control charts. Two of the most important are statistical design and economic design. In statistical design the objective is to have control charts signal shifts in the process quickly and accurately and to keep false signals to a minimum. Woodall (1985) has recently addressed the issue of statistical design.

The objective of economic design is choose a design that balances the costs of producing poor quality products and the costs of operating control charts. Duncan (1956) was one of the first to address the issue of economic design; for more recent work see, e.g. v. Collani (1989).

An alternative to statistical and economic design has been proposed by Saniga (1989) and is known as economic statistical design. Economic statistical design is a method in which statistical constraints are placed on economic models to yield designs that meet statistical requirements at minimum cost. Saniga (1989) has shown that these designs are an improvement to statistical designs since they achieve the same statistical goals but at minimum cost. And while they are more costly than economic designs they have other advantages such as guaranteeing high quality, keeping false searches at a minimum, and perhaps being robust to shifts other than those expected to occur. In this paper we investigate this issue of robustness in more detail. In this context we say a design is robust if its cost is smaller than other designs for shifts apart from the expected shift. Our results indicate that robustness is not a feature of any design in general, but that the flexibility of economic statistical design in achieving a goal of robustness makes it a preferred choice as a general method for designing control charts.

In the next section we discuss some of the advantages and disadvantages of the methods of control chart design. Section 3 describes an experiment in which the costs of economic designs, statistical designs and economical statistical designs are compared at shift levels apart from the expected shift. Finally, a summary is given in Section 4.

2. Design Methods

Three general methods of designing control charts are in use today. Without a doubt the most popular is Shewhart's heuristic, which is to take a sample of size four or five (for X and R charts), set three sigma control limits and sample perhaps once an hour. Shewhart's design is simple but is lacking in terms of its cost and effectiveness. Its popularity is due to its simplicity and familiarity.

Statistical design is perhaps as old as Shewhart's design. Statistical design guarantees high quality with the appropriate selection of ARL's and sampling frequency but can be unnecessarily costly.

The third method, economic design, is relatively new, beginning perhaps with the pioneering work of Duncan (1956) and carrying on to the present. The advantage of economic design is that all of the factors and costs that are measurable are considered in achieving a design. Thus the design is optimal in at least an economic sense. Some problems with economic designs have been noted by Woodall (1986) and include the possibility of a high Type I error probability, which implies a large number of false searches, something that production managers will not tolerate. Economic designs can also allow poor quality products if that policy is economically optimal. Deming (1982) has emphasized an opposite goal; that of achieving high quality as a necessary constraint in order to stay in business. Economic designs are also complex when compared to other design methods and it is perhaps this characteristic which leads to a lack of use in general in practice (Saniga and Shirland (1977).

Saniga (1989) has proposed another method of design, economical statistical design, which is related to both economic and statistical design. In economic statistical design constraints are placed on the economic model to ensure that false searches are kept at a low level and that assignable causes are detected quickly. In one sense this method is an improvement to pure statistical design since Saniga has shown that the same statistical properties are achieved but at minimum cost. In another sense this method modifies economic design to meet the requirements of production in a contemporary environment, requirements that have been outlined by Deming (1982). Since tighter control may be required than with economic design it is not unexpected that economic statistical designs will be more costly. But the tighter control should add to robustness in the sense that shifts apart from the expected shift will be detected more quickly and therefore less profit will be lost. Perhaps the most important attribute of economic statistical design is its flexibility. With the appropriate choice of design constraints the user can choose a purely statistical design, a purely economic design or a design meeting any of the temporal requirements of the system to which it is to be applied.

Economic statistical design is at least as difficult to implement as economic design since the same parameters must be estimated and more complex algorithms must be employed. Also, v. Collani (1989) has shown that economic statistical designs can allow the process to operate out of control more often than an economic design. Fortunately, if caution is used in the selection of the design constraints, this problem can be easily remedied as indicated in v. Collani, Saniga and Weigand (1989).

3. Experimental Results

We investigate the performance of economic designs, statistical designs and economic statistical designs when shifts apart from the expected shift occur. One must realize that the latter two design methods are unstructured in the sense that the designs are tailored to the problem depending upon the user's needs. Thus the comparison cannot be conclusive in finding the dominance of one method versus another. Rather, the purpose is to illustrate the costs that can be incurred if shifts other than expected shifts occur.

We consider cost or profit to be related to the magnitude of the shift in one of two forms, linear and quadratic. To explain in more detail let F be the cost per hour according to Chiu's (1975) economic model where

$$F = F(n, h, L_i)$$

$$\frac{\lambda M B_1 + T B_0 + \lambda W + (b + cn)(1 + \lambda B_1)/h}{1 + \lambda B_1 + t_0 B_0 + \lambda t_1}, \tag{1}$$

with

$$\tau = \{1 - (1 + \lambda h)\exp(-\lambda h)\}/[\lambda - \lambda \exp(-\lambda h)],$$

$$B_0 = \alpha (1 - \lambda \tau)/h,$$

$$B_1 = h/p - \tau,$$

$$M = V_0 - V_1,$$

$$T = A_0 + V_0 t_1,$$

and

$$W = A_1 + V_0 t_1.$$

The terms are: n – sample size, h – intersample interval, Li – control limit parameters, p – power, A_0 – the expected cost of a false alarm, A_1 – the expected cost of detecting and removing an assignable cause, V_0 – the profit per hour when the process is in control, V_1 – the profit per hour when a shift has occurred, b – the fixed cost of sampling, c – the variable cost of sampling, t_0 – the expected time to search after a false alarm, t_1 – the expected time to search and adjust for an assignable cause, and λ – the mean rate of occurrence of the assignable cause. We assume the process is in control if output is normally distributed with parameters μ_0 and σ_0. Further, we assume specifications of the product are set at μ_0 plus or 3 σ_0. Thus, when the process is in control the percent defective is d = .26%. We model the out of control profit function V_1 such that

$$V_1 = f(d) \tag{2}$$

and this function is linear or quadratic. In the linear case, the two points that define the function are V_0 yielding a percent defective of d = 0.26% and V_1 yielding a percent defective based upon the shift parameters μ_1 and σ_1 where μ_1 and σ_1 are the parameters of the process at the expected shift level. Note that with this model profits can be negative, a not unusual case in practice.

For the quadratic case we use the same points in determining the relationship in addition to a third point being a profit of V_0 at a percent defective of 0.00%.

Table 1a. compares the costs of economic design, statistical design and economic statistical design for a problem defined by Saniga (1989) at various levels of process parameters including those at the expected shift level.

At the expected shift level of $(\mu_1-\mu_0)/\sigma_0=1.5$, $\sigma_1/\sigma_0=2$ there is approximately a 25% savings in cost for using the economically optimal design, the economic design. For a smaller shift than expected the economic design is also preferred. But as the shift size gets larger the dominance of the economic statistical design becomes apparent, resulting in approximately a 23% savings for the largest shift when the profit function is linear. If we recalculate the economic statistical design with an unconstrained ATS (which, in effect allows the choice of the optimal ATS), we achieve the results shown in the last row of Table 1a. These results are very similar in cost to the economic design but have a much smaller Type I error probability $\alpha = 0.0052$ as compared to $\alpha = .0601$ for the econonmic design. The slight increase in cost coupled with the much smaller probability of a false search would yield a design more likely to be implemented in practice.

The statistical design in Table 1a. features a Type I error

$$\mu_1-\mu_0/\sigma_0, \quad \sigma_1/\sigma_0$$

		1,1	1.5,2	2,2	2.5,2	3,2.5
Economic – Statis-tical Design	L	1.80	2.31	2.63	3.14	3.93
	Q	1.87	2.31	2.43	2.60	2.80
Economic Design	L	1.05	1.94	2.55	3.52	4.95
	Q	1.15	1.94	2.19	2.50	2.88
Statisti-cal Design	L	6.38	6.51	6.60	6.73	6.94
	Q	6.39	6.51	6.54	6.58	6.37
Economic-Statistical Design**	L	1.03	1.97	2.60	3.61	5.11
	Q	1.15	1.97	2.22	2.55	2.95

Table 1a: Costs of Various Designs
for Selected Shifts*
(Expected shift $\mu_1-\mu_0/\sigma_0=1.5$, $\sigma_1/\sigma_0=2$)

$\alpha = 0.0052$ for all designs except economic design
$\alpha = 0.0406$ for economic design

L = Linear Profit Function

Q = Quadratic Profit Function

* $V_0 = 50$, $V_1 = 25$, $A_0 = 1$, $A_1 = 2$, $b = 5$, $c = {}_1 1$

$\lambda = .01$, $t_0 = 0.1$, $t_1 = 0.3$, $\mu_1-\mu_0/\sigma_0 = 1.5$, $\sigma_1/\sigma_0 = 2$

** Optimal ATS

probability of 0.0052 , power of 0.95 and an ATS of 1 hour. This tight design is relatively costly at all shift levels as indicated in the Table.

Table 1b. is a similar comparison to the last Table with the exception that the expected shift is $(\mu_1-\mu_0)/\sigma_0=1.$ and $\sigma_1/\sigma_0=1$. Again, the economic design is less expensive for expected shifts (which is always true) and those smaller. Note that economic design is also the most expensive design for the larger shifts and the difference is of practical significance; for the largest shift the difference is 91% larger than the cost of the economic statistical design and 546% larger than the cost of the statistical design. In fact the statistical design may be the design of choice at first glance, although an economic statistical design can always yield a design at least as good as statistical design as shown by Saniga (1989). To illustrate the flexibility of the economic statistical design we reduced the constraint bound on ATS to ATS<2; these results are presented in the last row of Table 1b. Note that the costs of this design indicate an alternative that may be compare favorably to the other designs. Also note that unless the probability of larger shifts were extremely small the economic design would not be a likely choice in practice.

It is of interest to compare cost results for all shifts in conjunction with the relative frequency of these shifts. That is, one may calculate the expected cost by finding the sum of the products of the probabilities of a shift times the costs of the shifts. This may lead to design based upon a multiple assignable cause model as investigated by a number of authors. In addition, one could investigate economic statistical designs under different constraint specifications to find a design that minimizes expected cost over the range of process shifts, or

$$\mu_1-\mu_0/\sigma_0, \quad \sigma_1/\sigma_0$$

	0.5,1	1,1	1.5,2	2,2	2.5,2.5
Economic - Statisti- cal Design L	2.16	2.52	57.65	96.65	182.55
Q	2.60	2.52	7.31	8.86	11.47
Economic Design L	1.23	2.02	109.12	184.02	349.78
Q	1.64	2.02	11.38	14.31	19.35
Statisti- cal Design L	7.23	7.31	21.78	32.21	54.96
Q	7.39	7.31	8.56	8.98	9.67
Economic- Statistical design** L	3.93	4.10	31.64	50.83	93.44
Q	4.17	4.10	6.49	7.24	8.54

Table 1b. Costs of Various Designs
for Selected Shifts*
(Expected shift $\mu_1-\mu_0/\sigma_0=1$, $\sigma_1/\sigma_0=1$)

$\alpha = 0.0052$ for all designs except economic design
$\alpha = 0.0601$ for economic design

* $V_0 = 50$, $V_1 = 25$, $A_0 = 1$, $A_1 = 2$, b = 5, c = .1

$\lambda = 0.01$, $t_0 = 0.1$, $t_1 = 0.3$

** ATS $\leq$ 2

build a multiple assignable cause model directly for the economic statistical design problem.

4. Summary

We present results that show the economic statistical design method can be more robust in situations in which shifts other than expected shifts occur. The flexibility of this method in developing alternative designs is also illustrated and it is argued that this flexibility is of importance in the context of firm wide decision making. These advantages coupled with other advantages as outlined by Saniga (1989) makes economic statistical design worthy of consideration as a general method for control chart design.

REFERENCES

[1] Chiu, W. K. (1975), "The Economic Design of Attribute Control Charts," Technometrics, 17, 81-87.

[2] v. Collani, E. (1989b), personal correspondence.

[3] v. Collani, E., Saniga, E. M. and C. Weigand, "Economic Adjustment Designs for X Charts," unpublished manuscript.

[4] v. Collani, E. (1989), The Economic Design of Control Charts, Teubner Verlag, Stuttgart.

[5] Deming, W. E. (1982), Quality, Productivity and Competitive Position, Cambridge, MA: MIT Press.

[6] Duncan, A. J. (1956), "The Economic Design of X Charts Used to Maintain Current Control of a Process," Journal of the American Statistical Association, 51, 228-242.

[7] Saniga, E. M. (1989), "Economic Statistical Control Chart Designs With an Application to X and R Charts," Technometrics, Vol. 31, No. 3, pp. 313-320.

[8] Saniga, E. M., and Shirland, L. (1977), "Quality Control in Practice: A Survey," Quality Progress, 10, 30-33.

[9] Shewhart, W. A. (1931), Economic Control of Quality of Manufactured Product, New York: Van Nostrand.

[10] Woodall, W. H. (1985), "The Statistical Design of Quality Control Charts," The Statistician, 34, 155-160.

[11] Woodall, W. H. (1986b), "Weaknesses of the Economic Design of Control Charts," Technometrics, 28, 408-409.

Approximately Optimal Economic Process Control for a General Class of Control Procedures

O. Hryniewicz, Warsaw, Poland

1. INTRODUCTION

In 1924 Shewhart introduced a new method for controlling the quality of a production process-the control chart. The most general control chart methodology consists of sampling from a process and evaluating the samples in order to find a signal that the considered production process is out-of-control. Whenever this state of the process is indicated searching and removing the assignable cause takes place. There are many propositions concerning the problem how to design control chart (see Lorenzen & Vance (1986) for references) - most of them are based rather on practical experience than on formal reasoning, reflecting the situation that up to now there does not exist a generally accepted method how to design control charts.

The problem of the optimal design of control charts using economic considerations was first solved by Duncan (1956). Since this very important work a great number of papers devoted to this problem have been published. Their results are summarized and reviewed in papers by Gibra (1975). Montgomery (1980) and Vance (1983).

Those models based on Duncan´s (1956) approach are unified in the paper of Lorenzen & Vance (1986). The objective function in this approach is the expected cost (loss) per hour. A modified approach was introduced by von Collani (1978,1981,1986) who uses as objective function the average profit per item produced in the long run. The advantage of this approach seems to result from significant reduction of the number of parameters entering explicitly the objective function.

In the majority of papers on economic desing of control charts exponential distribution of time between failutes is assumed. In the papers of Baker (1971). Heikes, Montgomery, and Yeung (1974), and Montgomery and Heikes (1976) non-Poisson models were assumed but only under the assumption that the system is renewed after each alarm,· true or false. Hu (1984) proposed an economic design of $\bar{x}$-chart for the Weibull model and a constant sampling interval.

Frontiers in Statistical Quality Control 4
Ed. by Lenz et al.
© Physica-Verlag Heidelberg 1992

Similar assumptions were also made in McWilliams (1989). The assumption of the constant sampling interval for the Weibull model was relaxed in Banerjee and Rahim (1988) who assumed a certain class of variable sampling intervals. Also in nearly all papers on economic design simple sampling procedures with constant sample size have been assumed. The only modification was due to sample curtailment as in Williams, Looney and Peters (1985).

In this paper the results of von Collani obtained for control charts of Shewhart-type (constant sample size) are generalized to the case of variable sample sizes (both attribute and variable, single - and multidimensional) under the assumption that there exists only one assignable cause for the process to go out of control. Moreover, it is assumed that the parameters which describe the process are known in both cases.i.e. for the in-control and the out-of-control state. Under this general framework different control procedures (including common control charts) can be investigated with respect to their economic performance.

2. MATHEMATICAL MODEL

We assume that the considered production process can be either in an acceptable STATE I characterized by a vector of parameters $p_1 = (p_{11}, p_{12}, \ldots, p_{1k})$, or in an unacceptable STATE II characterized by a vector of parameters $p_2 = (p_{21}, p_{22}, \ldots, p_{2k})$. The values of p_1 and p_2 are assumed to be known numerically. We further assume that the duration of STATE I is a random variable distributed accordingly to a known continuous distribution function $F(t)$. We assume that the moment of the transition from STATE I to STATE II is not directly observable and the transition from STATE II to STATE I can be achieved only by a special correction action. We consider three types of possible actions:

   a) monitoring of the process (sampling),
   b) inspection (searching for an assignable cause),
   c) renewal.

By monitoring we understand a statistical procedure which allows us to determine the actual state of the considered process with probability less than one. Thus, there are two types of error involved:

   Type I error means that the monitoring procedure will indicate the necessity of inspection while the process is operating in STATE I (probability of false alarm), and

Type II error means that the monitoring procedure will not indicate the necessity of inspection with a following renewal while the process is in the unacceptable STATE II.

We assume that the probabilities for a Type I error and for a Type II error exist and are given by certain functions $\alpha = \alpha(p_1, \gamma)$ and $\beta = \beta(p_2, \gamma)$ respectively, where $\gamma = (\gamma_1, \gamma_2, \ldots, \gamma_m)$ is a vector of parameters which fully describes the sampling procedure (e.g. for single sample np-chart: $\gamma_1 = n$, $\gamma_2 = c$). We assume that monitoring actions are performed periodically after h produced units, hence the expected number of minitoring actions while the process operates in STATE I is given by (see e.g. Duncan (1956))

$$E \; A_I \; = \; \sum_{i=1}^{\infty} R(ih) \tag{2.1}$$

where

$$R(ih) = 1 - F(ih) \tag{2.2}$$

The expected number of monitoring actions while the process remains in STATE II is given by

$$E\left[A_{II}\right] = \frac{1}{1 - \beta(p_2, \gamma)} \tag{2.3}$$

Hence, the expected time between consecutive renewals is given by

$$E\left[\tau_R\right] = \left(E\left[A_I\right] + E\left[A_{II}\right]\right) h \tag{2.4}$$

Next we introduce economic parameters in order to describe the economic consequences of the actions to be performed. The cost of a sampling action is described by three quantities:

a_0^* - fixed cost per sampling,

a_1^* - unit cost of sampling,

$n_i = n_i(p_i, \gamma)$ - the expected number of sampled elements, where $i=1$ in the case that the process is in STATE I during the sampling action, and $i=2$ in the case that it is in STATE II.

It is, of course, possible to introduce different unit costs of sampling taking a_{11}^* as a unit cost of sampling in STATE I, and a_{12}^* as a unit cost of sampling in STATE II. This difference, however, is in general of no significant importance for the optimal desing of a control chart.

Moreover we assume that the following costs are known:

e^* - average cost of an erroneous inspection during STATE I,

r^* - average cost of searching for an existing assignable cause and the following renewal,

g_1 - average profit derived from one unit produced while the process is operating in STATE I,

g_2 - average profit derived from one unit produced while the process is operating in STATE II.

For the model described above it has been shown in Hryniewicz (1988) that the optimal design of a control procedure (h^*, γ^*) for which the long term profit per unit produced is maximal can be found by maximization of the following objective function

$$G = \frac{g_1 E\left[\tau\right] + g_2 \{E\left[\tau_R\right] - E\left[\tau\right]\} - \alpha E\left[A_I\right] e^* - E\left[A_I\right] S_1^* - E\left[A_{II}\right] S_2^* - r^*}{E\left[\bar{\tau}_R\right]} \qquad (2.5)$$

where

$$S_1^* = a_0^* + n_1 a_1^* ,$$
$$S_2^* = a_0^* + n_2 a_2^* ,$$

and $E\left[\tau\right]$ is the expected duration of STATE I.

Function G can be transformed to the following objective function

$$G^*(h, \gamma) = e^* \left[\frac{b - E\left[A_I\right] \alpha - E\left[A_I\right] S_1 - E\left[A_{II}\right] S_2}{h \left(E\left[A_I\right] + E\left[A_{II}\right] \right)} \right] + g_2 \qquad (2.6)$$

where

$$S_1 = (a_0^* + a_1^* n_1)/e^* \qquad (2.7)$$

$$S_2 = (a_0^* + a_1^* n_2)/e^* \qquad (2.8)$$

and

$$b = \{(g_1 - g_2) E\left[\tau\right] - r^* \} / e^* > 0 \qquad (2.9)$$

where $E\left[\tau\right]$ is the expected duration of STATE I.

The quantities S_1 and S_2 have interpretation as the relative average costs of sampling actions in STATE I and STATE II, respectively. In a more general case a_1^* in (2.7) can be substituted by a_{11}^*, and a_1^* in (2.8) by a_{12}^*. Parameter b can be interpreted as the relative average benefit from one renewal of the process.

Obviously, the maximization of (2.6) is equivalent to the maximization of the following objective function:

$$G(h,\gamma) = \frac{G^*(h,\gamma)-g_2}{e^*} = \frac{1}{h}\ \frac{b-A_1\alpha-A_1S_1-A_2S_2}{A_1+A_2} \tag{2.10}$$

where

$$A_1 = E\left[A_I\right]$$
$$A_2 = E\left[A_{II}\right]$$

We notice that among the quantities entering (2.10) only A_1 is a function of the sampling interval h, i.e. $A_1=A_1(h)$.

In order to find the optimal value of the sampling interval h for a fixed sampling procedure γ we have to differentiate $G(h,\gamma)$ with respect to h and to equate this partial derivative to zero, leading to the following equation

$$-\frac{1}{h^2}\left[\frac{b-A_2S_2}{A_1+A_2}\ (\alpha+S_1)\ \frac{A_1}{A_1+A_2}\right]+\frac{1}{h}\left[-A_1'\ \frac{b-A_2S_2}{A_1+A_2}\ -\right.$$
$$\left.(\alpha+S_1)A_1'\ \frac{A_2}{(A_1+A_2)^2}\right] = 0 \tag{2.11}$$

Multiplying both sides of (2.11) by h^2, and making further transformations of (2.11) we obtain two equivalent equations

$$(S_2A_2-b)+\left[b+A_2(\alpha+S_1-S_2)\right]\phi_1(A_2,h)=0 \tag{2.12}$$

$$(\alpha+S_1)-\left[b+A_2(\alpha+S_1-S_2)\right]\phi_2(A_2,h)=0 \tag{2.13}$$

where

$$\phi_1(A_2,h) = \frac{A_1(A_1+A_2)-hA_1'A_2}{(A_1+A_2)^2} \tag{2.14}$$

$$\phi_2(A_2,h) = \frac{A_1+A_2+hA_1'}{(A_1+A_2)^2} \tag{2.15}$$

$$A_1' = \frac{\partial}{\partial h}\ A_1$$

When the duration of STATE I is exponentially distributed with $E[\tau]=1/\lambda$ it has been shown in Hryniewicz (1988) that in this case we have

$$\phi_1(A_2,h) = \frac{1+A_2\left[e^{\lambda h}(1+\lambda h)-1\right]}{\left[1+A_2(e^{\lambda h}-1)\right]^2} \tag{2.16}$$

The optimal sampling interval h^* can be obtained by solving numerically either (2.12) or (2.13), and the optimal sampling procedure γ^* can be found by numerical maximization of (2.6).

It is possible, however, to find a very good approximation to (h^*, γ^*) when the sampling interval h is small in comparison to $E(\tau)$. The approximately optimal control procedure can be obtained using the following theorem.

Theorem: For a sufficiently large value of b the optimal sampling interval h^* is given by

$$h^* = E(\tau) \sqrt{\frac{2(\alpha+S_1)}{(2A_2-1)\ b+A_2(\alpha+S_1-S_2)}} \qquad (2.17)$$

and the optimal sampling procedure γ^* can be found by minimization

$$\gamma^*: \quad \min_{\gamma} (2A_2-1)(\alpha+S_1) \qquad (2.18)$$

Proof:

Assume that the distribution F of τ is such that the following approximation holds for small h

$$F(x+h) \cong F(x)+f(x)h+0.5f(x)h^2 \qquad (2.19)$$

Hence

$$A_1(h) = \sum_{i=0}^{\infty} i\ F((i+1)h)-F(ih) \cong \sum_{i=0}^{\infty} ih\ f(ih)+0.5\sum_{i=0}^{\infty} ih^2 f'(ih) =$$

$$= \frac{1}{h}\sum_{i=0}^{\infty} (ih)f(ih)h+ 0.5 \sum_{i=0}^{\infty} (ih)f'(ih)h \qquad (2.20)$$

For small values of h the sums in (2.20) can be approximated by the corresponding integrals. Numerical experiments have revealed that in the case $h < 0.1E(\tau)$ these approximations are very accurate for a very wide class of distributions. Further information about the accuracy of similar approximations can be found in Cox (1957) and Stefansky & Kaiser (1973). Hence we have

$$A_1(h) \cong \frac{1}{h} \int_0^{\infty} uf(u)\,du+ 0.5 \int_0^{\infty} uf'(u)\,du = \frac{E(\tau)}{h} - 0.5 \qquad (2.21)$$

Taking $A_1'(h) \cong -E(\tau)/h^2$ and inserting both approximations into (2.16) we obtain

$$\phi_2(A_2,h) \cong \frac{(A_2-0.5)h^2}{\left[E(\tau)+(A_2-0.5)h\right]^2} \qquad (2.22)$$

Let $x=h/E(\tau)$ and expand $\phi_2(A_2,h)$ around 0 with respect to x, then for small values of x it follows

$$\phi_2(A_2,xE[\tau]) = (A_2-0.5)x^2-(2A_2-1)x^3+..=(A_2-0.5)x^2+o(x^3) \qquad (2.23)$$

Inserting (2.23) into (2.13) we arrive at a quadratic equation in h with a solution given by (2.17). It is worth to notice that exactly the same formula has been obtained in Hryniewicz (1988) for the special case of the exponential distribution starting from the exact formulae for $A_1(h)$ and $A_1'(h)$. This fact supports additionally our assertion that for relatively small values of h the approximation given by (2.22) is for practical reasons sufficiently accurate. From (2.17) it is obvious that for a sufficiently large value of b we have a small value of a sampling interval h, and therefore the assumption the approximation (2.22) is based on holds. This ends the proof of the first part of the theorem.

Consider now the problem how to determine the optimal sampling procedure $\gamma = \gamma(\gamma_1,\ldots,\gamma_m)$ when the sampling interval h is given. The objective function (2.10) can be expressed as follows

$$G(h^*,\gamma) = \frac{1}{h^*}\left[\frac{b-A_2S_2}{A_1^*+A_2} - (\alpha+S_1)\frac{A_1^*}{A_1^*+A_2}\right] \qquad (2.24)$$

where $A_1^*=A_1(h^*)$.

Now notice that for h^* (2.12) and (2.13) hold. Hence we can calculate the values of $b-A_2S_2$ and $\alpha+S_1$ from (2.12) and (2.13), respectively, and insert them to (2.24) arriving after some transformations at

$$G(h^*,\gamma) = -\frac{A_1'(h^*)}{(A_1^*+A_2)^2}\left[b+A_2(\alpha+S_1-S_2)\right] \qquad (2.25)$$

Next, we can rewrite (2.12) as

$$\frac{b+S_2A_2}{b+A_2(\alpha+S_1-S_2)} - \frac{A_1(h^*)}{A_1(h^*)+A_2} = -h^*A_2\frac{A_1'(h^*)}{(A_1(h^*)+A_2)^2} \qquad (2.26)$$

We notice that for small h^* we approximately have

$$\frac{A_1(h^*)}{A_1(h^*)+A_2} \cong 1 \qquad (2.27)$$

Hence,

$$- \frac{A_1'(h^*)}{(A_1(h^*)+A_2)^2} \approx - \frac{1}{h^*} \frac{\alpha+S_1}{b+A_2(\alpha+S_1+S_2)} \tag{2.28}$$

and, finally, the optimal sampling procedure can be determined by minimization of

$$\tilde{G}^* = \frac{\alpha+S_1}{h^*} \tag{2.29}$$

Suppose that the value of b is large enough that $b \gg A_2(\alpha+S_1-S_2)$ and (2.17) holds. Inserting h^* from (2.17) into (2.29) we find that minimization of (2.29) is equivalent to minimization of (2.18) which completes the proof.

The objective function $\tilde{G}^*$ has a natural interpretation. From (2.17) we can see that

$$\alpha+S_1 \sim (h^*)^2 (A_2-0.5) \tag{2.30}$$

Hence,

$$\tilde{G}^* \sim h^* (A_2-0.5) \tag{2.31}$$

i.e. $\tilde{G}^*$ is approximately proportional to the expected number of elements produced while the process operates in the unacceptable state.

3. PROPERTIES OF THE OPTIMAL PROCEDURES. NUMERICAL EXPERIMENTS.

The theoretical results of the previous section will be illustrated by some numerical examples. First, we will investigate the difference between optimal and approximately optimal solutions. For the np-control charts it has been shown in Hryniewicz (1988) that in the exponential case approximately optimal solutions are generally very close to exact ones. In the case of an arbitrary distribution F(t) of the duration of STATE I the only possible reason for the worsening of this accuracy is the approximation (2.21) of $A_1(h)$.

This approximation, however, seems to be very accurate for a wide class of distributions and a wide range of input parameters. Now let us consider a similar example for the $\bar{x}$-control chart characterized by the following parameters: n - sample size, k - upper limit (in standard deviations, and h - sampling interval (in units produced).

Assume the exponential distribution of the time between consecutive disorders of the production process and the following set of parameters:

$r^* = 100.$, $e^* = 50.$, $a_0^* = 0.$ $g_1 = .5$ $g_2 = .37$

Denote by δ the shift of the mean value of the process from the target value in STATE I (measured in standard deviations) and by λ the hazard rate of the life-time distribution $F(t)$. In Table 1 the comparison of the optimal and approximately optimal $\bar{x}$-control chart is presented for different values of δ, unit sampling cost a_1^*, and the relative average benefit from one renewal of the process b.

Parameters	b	Optimal $\bar{x}$-chart	Appr.opt.$\bar{x}$-chart
$\delta = 1.$ $a_1^* = 0.005$ $\lambda = 0.00022$	10	$n = 22$ $k = 3.472$ $h = 91.8$ $G^* = .475313$	$n = 22$ $k = 3.476$ $h = 90.8$ $G^* = .475313$
$\delta = 0.5$ $a_1^* = 0.05$ $\lambda = 0.00005$	50	$n = 44$ $k = 2.310$ $h = 815.3$ $G^* = .488273$	$n = 45$ $k = 2.328$ $h = 803.9$ $G^* = .488270$
$\delta = 1.5$ $a_1^* = 0.05$ $\lambda = 0.00005$	50	$n = 8$ $k = 3.059$ $h = 341.8$ $G^* = .492327$	$n = 8$ $k = 3.062$ $h = 338.6$ $G^* = .492337$

Table 1: Optimal and approximately optimal $\bar{x}$-charts.

The differences between optimal and approximately optimal plans are of any practical importance even for relatively small values of b. For the values of b greater than 50 they arise rather from rounding up errors in optimization routines than from the differences in objective functions. It is worthwhile to notice that in the considered examples parameter k is often far from values recommended by many authors (i.e.from 3). This happens, for example, when the expected shift from the target value δ is rather small (e.g. less than 1).

Now we can compare different sampling procedures when they are used for the same production process. To illustrate this problem we consider the following production process. Assume that the process to be controlled is characterized by the following pra- meters:

production rate v = 200 units/hour,

profit from one conforming item c_1 = 20,

loss from one nonconforming item c_2 = -100,

and, $a_0^*=0.$, $a_1^*=1.$, $r^*=1000.$, $e^*=500.$, $\lambda =0.00005$ 1/unit produced .

Moreover assume that quality characteristic to be controlled is normally distributed with parameters $\mu_1=100.$, and $\sigma=2$. Suppose that the considered process deteriorates when its mean value is shifted to a new value $\mu_2=100.3$ (σ remains unchanged).

Let $\mu_0=100.$ be a target value of the considered quality characte- ristic, and L=99.5 U=100.5 be its lower and upper tolerance limits, respectively. Hence, we have:

fraction nonconforming in STATE I: $p_1=0.0124$,

fraction nonconforming in STATE II: $p_2=0.2$.

Now we can calculate

$$g_1 = c_1*(1-p_1)+c_2*p_1 =18.51,$$
$$g_2 = c_1*(1-p_2)+c_2*p_2 = -4.$$

To compare different control procedures we calculate the following characteristics:

Loss of efficiency: $L_e = G_i-G$,

where $G_i=g_1- \lambda *r^*$ is the profit per item produced in the ideal case of constant monitoring (without additional costs) of the considered process,

Probability of false alarm in STATE I: α ,

Precentage of time in STATE II:

$$P = \frac{\text{duration of STATE II}}{\text{length of renewal cycle}} \cdot * 100\%,$$

Sampling effort:

$$S_e= \frac{\text{number of sampled}}{\text{length of renewal cycle}} * 100\%$$

The results of the comparison of different procedures (both attri- butes and variables) are presented in Table 2.

For the optimal procedures considered sampling intervals h^* have
been rounded up to values more convenient from operational reasons.
The exact values of h^* sre given in brackets.

Sampling plan	Parameters	h^*	L_e	α	P	S_e
variables sing. (opt.$\bar{x}$-chart)	$n = 7$ $k = 2.85$	100 (105)	0.15	0.002	0.34	6.63
variables sequential (Wald´s)	$h_1=-0.344$ $h_2=7.844$ $s^2=0.75$	30 (33)	0.07	<0.001	0.15	4.11
variables sing. (trad.$\bar{x}$-chart)	$n = 5$ $k = 3$	200	0.27	0.001	1.06	2.50
attributes sing. (opt.np-chart)	$n = 11$ $c = 1$	100 (116)	0.26	0.007	0.49	11.0
attributes double	$n_1=n_2=10$ $a_1 = 0$ $r_1=r_2=3$	100 (114)	0.21	0.001	0.46	9.8
attributes sequential (Wald´s)	$h_1=-0.048$ $h_2= 3.23$ $s\leq0.07$	15	0.16	<0.001	0.36	8.12

Table 2: Camparison of different control procedures.

4. DISCUSSION

The main advantages of the model can be summarized as follows.

a)   Quality characteristics to be controlled may be discrete
(e.g. attributes) or continuous (variables).

b)   Quality characteristics to be controlled may be one-dimen-
sional (e.g. the mean value of a single variable) or multi-dimen-
sional (e.g. both the mean value and the standard deviation),
(another approach is considered by Arnold (1988)).

c)   The statistical procedure adopted for the detection of
the process deterioration is not limited to single sampling plans.

d)   There is no restriction to the distribution of times between consecutive deteriorations of the production process.

e) There exist approximate solutions to the original optimization problems which dramatically simplify the optimization procedure while the results of optimization remain very accurate within a wide range of input parameters.

Additional positive feature of the proposed procedures can be derived from the analysis of (2.18). It is obvious the these procedures should have small probabilities of Type I error α and a possibly short average run length while the process remains in an unacceptable STATE II.

The most important possible restrictions for the application of the proposed approximate solution are the following.

a) Sampling costs are so small in comparison with a cost of erroneous inspection (false alarm) that the average number of sampled elements can be compared with the average number of elements produced between consecutive inspections. In this case the model adopted in this paper is not valid anymore. The discussion of this problem can be found in Hryniewicz (1988).

b) Distribution of τ has sharp peaks or discontinuities of its density function. In this case the approximation of $A_1(h)$ may not be accurate.

The analysis of the comparison of different control procedures can be summarized as follows:

a) procedures based on sequential sampling plans are more efficient than procedures based on single sampling plans (control charts).

b) optimal procedures based on variables sampling plans are generally more efficient than procedures based on attributes sampling plans, but when unit sampling costs for attributes sampling plans are smaller than in the case of variables sampling this advantage may not exist anymore,

c) traditional $\bar{x}$-control charts may be less efficient even than optimal np-control charts,

d) traditional np-control charts with the requirement that both control limits exist in many cases have no practical sense (e.g. in the considered case required sample size is close to 800),

213

e) statistical properties of optimal procedures measured by
and P are good, and especially in the case of sequential procedures are much better than the properties of traditional sampling procedures based on "statistical" approach.

APPENDIX: Symbols used in the economic design of control procedures.

Input variables:

g_1 - profit per produced element while the process is in control.

g_2 - profit per produced element while the process is out of control.

λ - 1/mean time process is in control.

a_0^* - fixed cost per sampling action.

a_1^* - cost per unit sampled.

e^* - cost of search action (false alarm).

r^* - cost of search and renewal action.

P_1 - vector of parameters characterizing the process in control.

P_2 - vector of parameters characterizing the process out of control.

δ - shift of the mean value of a parameter characterizing the
    quality of the production process (measured in standard devia-
    tions).

Output (optimized) parameters:

h - sampling interval (in produced elements).

γ - vector describing the sampling procedure (e.g. for a np-chart:
$\gamma = (\gamma_1, \gamma_2)$, where $\gamma_1 = n, \gamma_2 = c$; for a procedure based on Wald's
    sequential sampling plan: $\gamma = (\gamma_1, \gamma_2, \gamma_3)$, where $\gamma_1 = h_0$, $\gamma_2 = h_1$
    and $\gamma_3 = s_0$; for $\bar{x}$-chart $\gamma = (\gamma_1, \gamma_2)$, where $\gamma_1 = n, \gamma_2 = k$).

Calculated variables:

$G^*(h, \gamma)$ - relative profit per produced item (the objective
function.

α - probability of false alarm.

b - relative profit per renewal.

A_1 - expected number of sampling actions while the process is in
control.

A_2 - expected number of sampling actions while the process is out
of control.

S_1 - relative expected cost of sampling action while the process
is in control.

S_2 - relative expected cost of sampling action while the process is out of control.

Acknowledgements: The main results of the paper were obtained during the author´s DFG scholarship in Wuerzburg, Germany. The author is indebted to Prof. Elart von Collani for the introduction to the subject and many helpful discussions.

5. REFERENCES

[1]   ARNOLD, B (1988): Optimal Control of a Process with Several components, Techn.Reports of the Wuerzburg Research Group on Quality Control, No.14, Univ. of Wuerzburg.

[2]   BAKER, K.R. (1971): Two Process Models in the Economic Design of an $\bar{x}$ Chart, AIIE Transactions, Vol.3, 257-263.

[3]   BANERJEE, P.K., and RAHIM, M.A. (1988): Economic Desing of $\bar{x}$-Control Charts Under Weibull Shock Models, Technometrics, Vol. 30, 407-414.

[4]   VON COLLANI, E. (1978): Kostenoptimale Pruefplaene fur die laufende Kontrolle einess normalverteilten Merkmals, Thesis, Wuerzburg.

[5]   VON COLLANI, E. (1981): Kostenoptimale Pruefplaene fuer die laufende Kontrolle eines normalverteilten Merkmals, Metrika , Vol. 28, 211-236.

[6]   VON COLLANI, E. (1986): A Simple Procedure to Determine the Economic Design of an $\bar{X}$-Control Chart, Journal of Quality Technology, Vol.18, 145-151.

[7]   VON COLLANI, E. (1989): The Economic Design of Control Charts. Stuttgart: B.G.Teubner.

[8]   COX, D.R. (1957): Note on Grouping, Journal of the American Statistical Society, Vol. 52, 543-547.

[9]   DUNCAN, A.J. (1956): The Economic Design of $\bar{X}$-Charts Used to Maintain Current Control of a Process, Journal of the American Statistical Association, Vol. 51, 228-242.

[10]   HEIKES, R.G., MONTGOMERY, D.C., and YEUNG, J.Y.H. (1974):
       Alternative Process Models in Economic Design of
       T^2 Control Charts, <u>AIIE Transaction</u>,Vol. 6, 55-61.

[11]   HRYNIEWICZ, O. (1988): The Economic Design of a Certain
       Class of Control Charts: A General Approach, <u>Techn.</u>
       <u>Reports of the Wuerzburg Research Group on Quality</u>
       <u>Control</u>, No.11, Univ. of Wuerzburg.

[12]   HU, P.W. (1984): Economic Design of an $\bar{x}$-Control Chart
       Under Non-Poisson Process Shift, Abstract, TIMS/ORSA
       Joint National Meeting, San Francisco, May 14-16,
       p.87.

[13]   LORENZEN, T.J., and VANCE, L.C. (1986): The Economic
       Design of Control Charts: A Unified Approach,
       Technometrics, Vol. 28, 3-10.

[14]   McWILLIAMS, T.P. (1989): Economic Control Chart Designs
       and the In-Control Time Distribution: A Sensitivity
       Study, <u>Journal of Quality Technology</u>,Vol. 21,103-110.

[15]   MONTGOMERY, D.C. (1980): The Economic Design of Control
       Charts: A Review and Literature Survey, <u>Journal of</u>
       <u>Quality Technology</u>, Vol. 12, 75-87.

[16]   MONTGOMERY, D.C. and HEIKES, R.G. (1976) Process Failure
       Mechanism and Optimal Design of Fraction Defective
       Control Charts, <u>AIIE Transactions</u>, Vol. 8, 467-473.

[17]   STEFANSKY, W. and KAISER, H.F. (1973): Note on Discrete
       Approximations, <u>Journal of the American Statistical</u>
       Association, Vol. 68, 232-234.

[18]   VANCE, L.C. (1983): A Bibliography of Statistical Quality
       Control Chart Techniques, 1970-1980, <u>Journal of Quality</u>
       <u>Technology,</u> Vol. 15, 59-62.

[19]   WILLIAMS, W.W, LOONEY, S.W. and PETERS, M.H. (1985):
       Use of Curtailed Sampling Plans in the Economic Design
       of np-Control Charts, <u>Technometrics</u>, Vol.
       27, 57-63.

Control Charts for Environmental Data

H. Schneider, Y. Hui, and J.M. Pruett, Baton Rouge, Louisiana, USA

1. INTRODUCTION

The ability to accurately monitor the level of toxic pollutants present in the environment has become a problem of great concern to government, business, the public, and to researchers. Recently, the EPA suggested using control charts to monitor toxic pollutants in the environment. However, the reporting procedures used are complicated by at least three factors. First, the sample sizes used to characterize the mean level of a toxic contaminant are typically small. The primary reason for taking small samples is simply a matter of technology and economics, i.e., measurements of toxic contaminants usually involve expensive laboratory tests. As a result, the estimated mean value of a given pollutant is based only on a small sample of observations. Even so, when the underlying data from which the sample has been drawn were to follow a normal distribution, the usual small-sample confidence intervals are often very wide.

The second complicating factor is that the precision of the instrumentation available has sometimes hampered the pollutant monitoring process. Because many instruments can only detect the pollutants present above some detection limit, it is not possible to know the values of the concentrations below the detection limit. For example, the instrumentation may show a zero parts per million (PPM) pollution reading when pollutants are actually present in the sample but below the instrument's ability to detect the pollutants. In effect, the sample of environmental data is often censored. This censoring makes it difficult for realistic estimates to be made of population parameters.

The third complicating factor is that for most environmental data, the assumption that the distribution is normal is not true. Instead, toxic pollutant concentrations often show a skewed distribution pattern. In other words, toxic pollutants tend to have more extreme observations than one would expect under normality. If the distribution of environmental data is not normal, the usual procedure for estimating the sample mean will not provide a good estimate of the population mean. Fortunately, the situation can be improved by transforming the data so that it appears to be normally distributed. Traditionally, a logarithmic transformation has been applied to pollutant data, leading to an estimation procedure based on the lognormal distribution. A general class of power transformations attributed to Box and Cox (1964) are applied in this article. Although the use of logarithmic transformations for purposes of stabilizing variances and transforming data to approximately normal form is well documented in the literature, there may be occasions in which other transformations or perhaps even no transformation at all may be more appropriate.

The problem discussed in this article deals with control charts for censored, nonnormal, environmental data. To establish an appropriate control chart, procedures for estimating the center line and control limits based on estimates of the mean and standard deviation from censored, nonnormal data are discussed. The problem is examined from two major perspectives. First, the questions of whether or not it is necessary to transform the data values before plotting them and, if a transformation is appropriate, what transformation should be made are addressed. The aforementioned class of Box-Cox power transformations is employed when a transformation is called for. Second, the question of how censored data values affect the performance of the control chart is examined. We will use the maximum likelihood estimator (MLE) approach on the censored, transformed data to find our estimator for the mean of the transformed distribution. Maximum likelihood estimators applied to censored data problems were discussed earlier by Shumway, Azari, and Johnson (1989), Cohen (1988), and Schneider (1986).

2. DATA TRANSFORMATION

In the analysis of data, it is often assumed that the observations $Y_1, Y_2, \ldots, Y_n$ are independent and normally distributed with constant variance. However, in reality, this assumption is sometimes violated. For example, the following processes may produce skewed distributions:

(1) Purity Tests: When testing the purity of products, samples often show a left-skewed distribution. The closer a product is able to come to 100% purity, the more skewed the data tend to become.

(2) Parts per Million Counts: When measuring some minor characteristic by counting the number of parts per million present, counts are frequently near zero.

If the data do not adhere to the independent and normally distributed assumption, transformations may be applied to improve matters. In this article, the general class of power transformations developed by Box and Cox (1964) are applied. Box and Cox developed an approach in which they used a transformation that simultaneously makes the model linear in the parameters, the variance homogeneous, and the transformed data distribution normal.

The Box and Cox method works with a parametric family of transformations from y to $y^{(\lambda)}$, with the parameter λ defining a particular transformation. The general transformation is

$$Y^{(\lambda)} = \begin{cases} (y^{\lambda}-1)/\lambda, & \lambda \neq 0, \\ \ln(y), & \lambda = 0. \end{cases} \qquad (1)$$

The transformation (1) holds for $y > 0$ and assures that, under weak conditions, the transformed observations are approximately normally distributed and the variance is constant (i.e., does not depend on the level of y). We will use the maximum likelihood approach to estimate the value of λ.

Using the Box and Cox power transformation, the original data may be transformed. To set the notation, assume that independent observations are available of which n_u are observed and n_l are below the measurement instrument's detection limits (i.e., censored data). Suppose the observed values are denoted by $y_1, y_2, \ldots, y_N$ and that, if observation y_i is censored, we know only that $y_i \leq D$, where D is some lower detection limit.

The transformed variables x_i are then defined as:

$$x_i = \begin{cases} (y_i^{\lambda}-1)/\lambda, & \lambda \neq 0, \\ \ln y_i, & \lambda = 0, \end{cases} \qquad (2)$$

where y_i are the observed values and the transformed threshold is

$$D^{\bullet} = \begin{cases} (D^{\lambda}-1)/\lambda, & \lambda \neq 0, \\ \ln D, & \lambda = 0. \end{cases} \qquad (3)$$

The log likelihood function of the original (censored) observations $y_1, y_2, \ldots, y_N$ and the transformed observations $x_1, x_2, \ldots, x_N$ becomes

$$\ln L(\lambda,\mu,\sigma) = -n_u \ln\sigma - \sum_{i=1}^{n_u} (x_i-\mu)^2/\sigma^2 + n_l \ln\Phi[(D^*-\mu)/\sigma]$$
$$+ (\lambda-1)\sum_{i=1}^{n_u} \ln(y_i) \tag{4}$$

For fixed λ, the last term is a constant and the problem is reduced to a single Type I censor sample. (See Schneider (1986) for a more detailed description of censored data.)

We will use a two-step procedure to find the optimal λ value. First, we will find a series of MLEs μ and σ, given several fixed values of λ. The log likelihood function $\ln L(\lambda,\mu,\sigma)$ is then evaluated for this series of λ values to determine the optimal λ.

The likelihood equations for fixed λ are

$$\frac{\partial \ln L}{\partial \mu} = -n_l \frac{\phi(u_l)}{\Phi(u_l)} \frac{1}{\sigma} + \sum_{i=1}^{n_u} \frac{(x_i-\mu)^2}{\sigma^2} \tag{5}$$

and

$$\frac{\partial \ln L}{\partial \sigma} = -n_l \frac{\phi(u_l)}{\Phi(u_l)} \frac{u_l}{\sigma} - \frac{n_u}{\sigma} + \sum_{i=1}^{n_u} \frac{(x_i-\mu)^2}{\sigma^2} \tag{6}$$

where $u_l = (D^*-\mu)/\sigma$

By letting $h = n_l/N$ and $W(u) = \phi(u)/\Phi(u)$, we may then rewrite the above likelihood equations in the following form shown below to obtain Cohen's (1959) maximum likelihood estimators. That is,

$$(\mu-\overline{x})/\sigma - [h/(1-h)]W(u_l) = 0 \tag{7}$$

$$s^2 + (\overline{x}-\mu)^2 - \sigma^2\{1-u_l[h/(1-h)]W(u_l)\} = 0 \tag{8}$$

where

$$\overline{x} = \sum_{i=1}^{n_u} x_i/n_u \tag{9}$$

$$s^2 = \sum_{i=1}^{n_u} (x_i-\overline{x})^2/n_u \tag{10}$$

are the mean and variance of the transformed values, respectively. By letting

$$Y(h,u_l) = [h/(1-h)]W(u_l), \tag{11}$$

we then obtain Cohen's equations for the MLEs. That is,

$$\sigma_{ML}{}^2 = s^2 + g(h,u_l)(D^*-\overline{x})^2 \tag{12}$$

$$\mu_{ML} = \overline{x} + g(h,u_l)(D^*-\overline{x}) \tag{13}$$

where

$$g(h,u_l) = \frac{Y(h,u_l)}{Y(h,u_l) + u_l} \tag{14}$$

and u_l is the solution to

$$\frac{1 - Y(h,u_l)[Y(h,u_l)-u_l]}{[Y(h,u_l)-u_l]^2} = \frac{s^2}{(D^*-\overline{x})^2} \tag{15}$$

The parameter value λ^*, which maximizes the log-likelihood function (4), may be used to determine the transformation of the data. The MLE μ_{ML} can then be used as the center line for the control chart.

3. ESTIMATION OF THE TRANSFORMED DATA'S MEAN FOR SMALL SAMPLES

After a transformation has been performed based on m samples of size n so that N=mn, a center line and a set of control limits for the transformed data may be computed. The center line is given by the MLE μ_{ML}. To determine control limits, the large sample variance of the MLE could be used. This method, however, is unreliable for small sample sizes. Hence, we suggest the use of the computed variance based on m MLE estimates taken from samples of size n. In this section, we describe the MLE estimate for known σ. The known variance MLE should be used when the process is in control and an estimate of σ is available from a large set of data.

By letting

$$\delta = (\hat{\mu}-\overline{x})/\sigma \tag{16}$$

and

$$\theta = (D^*-\overline{x})/\sigma, \tag{17}$$

then the MLE for known σ can be found by solving

$$\delta = Y(h,\theta-\delta) \tag{18}$$

for δ, where $h = n_l/n$. The maximum likelihood estimator of the mean is then

$$\hat{\mu} = \overline{x} + \delta\sigma. \tag{19}$$

In the next section, we demonstrate the method using an example.

4. EXAMPLE

Sixteen samples, each containing five measurements, show a measure of the pollutants present in air quality samples. The measurements are recorded in parts per million (PPM). The data are given in Table 1.

Sample Number	Sample Values				
1	29	30	12	15	28
2	20	24	8	0	16
3	22	20	21	0	11
4	11	25	19	22	10
5	10	13	0	19	12
6	14	9	18	32	14
7	0	0	13	12	25
8	14	26	9	0	8
9	22	10	17	14	13
10	8	23	11	35	39
11	13	14	14	33	0
12	9	14	13	11	13
13	17	13	0	8	10
14	18	35	14	27	13
15	20	10	0	14	0
16	16	12	16	12	21

Table 1: Measure of Pollutants in Air Quality Samples (in PPM)

The detection limit for the instrument used to make the measurements was 8 PPM. Hence, measurements below 8 are censored. They are shown in Table 1 as zeros.

Based on the Table 1 data set and its censoring point, we will estimate the center line and control limits using the procedures described in sections 2 and 3. The first step is to determine whether or not the process data are normally distributed. If not, the aforementioned power transformation is employed on the observed data. We then evaluate the log-likelihood function ln $L(\lambda, \mu, \sigma)$ to create a grid of λ values and determine the optimal λ^*, after which the MLEs $\hat{\sigma}$ and $\hat{\mu}$ based on the transformed data are found.

Based on examinations of the histogram and normal probability plot shown in Figures 1 and 2, it is clear that the process data are skewed. Hence, the data should be transformed.

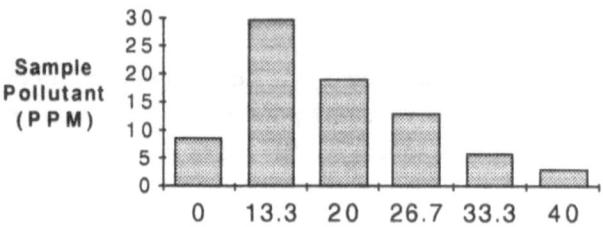

FIGURE 1: Histogram of the Number of Pollutants in Air Quality Samples

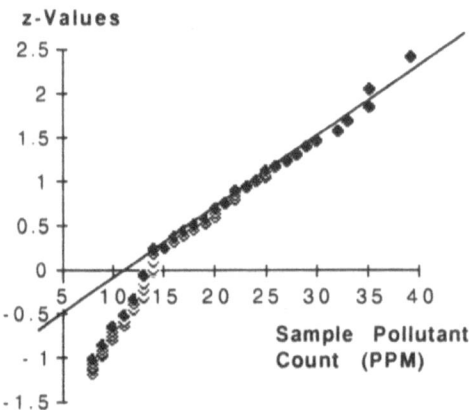

FIGURE 2: Normal Probability Plot of the Number of
Pollutants in Air Quality Samples

By considering λ values in the range $-0.6 < \lambda < 0.6$, we obtain a variety of MLEs $\hat{\mu}$ and $\hat{\sigma}$ over that interval. For each λ, the value of the maximum likelihood function as a function of λ is calculated. These are shown in Figure 3. The maximum log-likelihood function value is obtained at $\lambda = 0$. The corresponding MLEs for $\hat{\mu}$ and $\hat{\sigma}$ are 2.6459 and 0.4794, respectively.

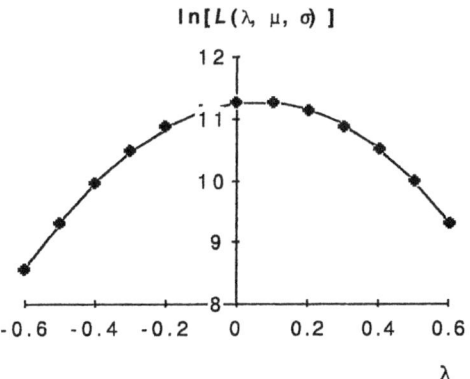

FIGURE 3: Log-Likelihood Functions

A normal probability plot of the log-transformed data is shown in Figure 4. It can be seen that the log data follow a line that is approximately straight, indicative that the transformed data is approximately normal.

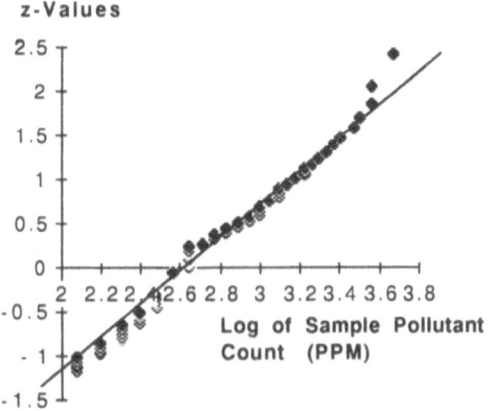

FIGURE 4: Normal Probability Plot of Log-Transformed Data

The MLE $\hat{\mu}$ = 2.6459 of the transformed data for λ=0 is chosen for use as the center line on the control chart. We use the 16 samples of size 5 each to obtain individual MLEs of the mean where the standard deviation of the 80 values (i.e., $\hat{\sigma}$=0.4794) is used in equations (16) to (19). The 16 MLEs are listed in Table 2.

Sample Number	MLE $\hat{\mu}$
1	2.9875
2	2.3034
3	2.3854
4	3.0847
5	2.6228
6	2.5123
7	2.6510
8	2.8618
9	2.3714
10	2.4183
11	2.8865
12	2.8488
13	2.3366
14	2.5975
15	2.6323
16	2.7392

Table 2: 16 Sample MLEs

After computing the 16 MLEs, we estimate the standard error of the estimate as:

$$\sqrt{[1/16 \ \Sigma(\hat{\mu}_i - \hat{\mu})^2]} = 0.2439$$

The control limits are set three standard errors above and below the center line. If other limits are desired, they may be set using z-values from the standard normal distribution table. The numerical values of the center line and control limits for this example are:

Center Line   = 2.6459

Upper Control Limit   = 2.6459 + 3(0.2439) = 3.3776

Lower Control Limit   = 2.6459 − 3(0.2439) = 1.9142

The control chart is shown in Figure 5. In this case, it appears that the process is in control.

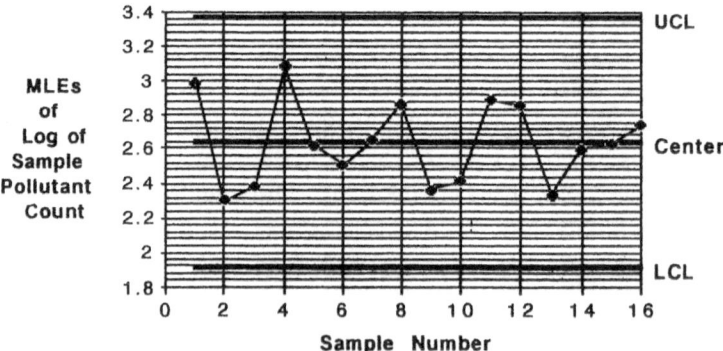

FIGURE 5: Control Chart for MLEs of Log-Transformed Data

5. SUMMARY

In this paper, we discussed an approach for developing control charts for censored, nonnormal, environmental data. The control chart development process for the data may be summarized in the following steps:

Step 1: Obtain m samples of size n each.

Step 2: Determine whether or not the process data are normally distributed and, assuming they are not, transform them using the Box and Cox procedure.

Step 3: Evaluate the log-likelihood function $\ln L(\lambda, \mu, \sigma)$ for a range of λ values between -1 and +1 and choose the optimal (i.e., maximum) λ value, λ^*.

Step 4: Using λ^*, calculate the MLEs $\hat{\mu}$ and $\hat{\sigma}$ using the procedure described previously.

Step 5: Compute MLE $\hat{\mu}_i$ for each sample, i.e., $i=1,2,\ldots,m$, using the standard deviation estimated in step 4.

Step 6: Using the m MLEs $\hat{\mu}_i$ computed in Step 5, estimate the standard error of $\hat{\mu}$, $\hat{\sigma}_\mu$.

Step 7: Finally, using the $\hat{\mu}$ and $\hat{\sigma}_\mu$ found in prior steps, compute appropriate control limits $\hat{\mu} \pm Z_{1-\alpha/2} \hat{\sigma}_\mu$ and develop the required control charts.

REFERENCES

[1] Box, G.E.P., and Cox, D.R. (1964). "An Analysis of Transformations" (with discussion), Journal of the Royal Statistical Society, Ser. B, Vol. 39, pp. 211-252.

[2] Cohen, A.C. (1959). "Simplified Estimators for the Normal Distribution When Samples Are Singly Censored or Truncated," Technometrics, Vol. 1, pp. 217-237.

[3] Cox, D.R., and Hinkley, D.V. (1974). Theoretical Statistics. London, England. Chapman & Hall.

[4] Schneider, H. (1985). "The Performance of Variable Sampling When the Normal Distribution Is Truncated," Journal of Quality Technology, Vol. 17, No. 2, 74-78.

[5] --- (1986). Truncated and Censored Samples from Normal Population. Marcel Dekker, Inc. New York, NY.

[6] Johnson, N.L., and Kotz, S. (1970). Continuous Univariate Distributions. John Wiley & Son, Inc. New York, NY.

[7] Hinkley, D.V., and Runger, G. (1984). "The Analysis of Transformed Data" (with discussion), Journal of the American Statistical Association, Vol. 79, pp. 302-320.

226

[8] Shumway, R.H., Azari, A.S., and Johnson, P. (1989). "Estimating
     Mean Concentrations Under Transformation for Environmental Data
     with Detection Limits," Technometrics, Vol. 31, No. 3, pp.
     347-356.

[9] United States Environmental Protection Agency, "Measure of Air
     Quality: The New Pollutant Standard Index," July 1978, OPA
     11/8.

Statistical Process Control for the Process Industries

G. B. Wetherill, Newcastle upon Tyne, England
D.W. Brown, Cleveland, England

1. INTRODUCTION

Our objective in this paper is to give some emphasis to the special problems of the process industries. Most of the literature on SPC relates to the component manufacturing field, and this has some unfortunate consequences, not the least of which is that the rather difficult problems of the process industries are not getting the interest they deserve.

2. SPC BACKGROUND

2.1 WHAT IS SPC?

Statistical process control and allied techniques of sampling inspection and quality control were developed in the 1920's. In May 1924, Walter A. Shewhart of Bell Telephone Laboratories developed the first sketch of a modern control chart. Work by him, H.F. Dodge, H.G. Romig, W.J. Jennett and others continued apace. In 1931 a crucial paper on the new techniques was presented to the Royal Statistical Society, which stimulated interest in the UK.

SPC was used extensively in World War II both in the UK and in the USA, but lost its importance as industries converted to peacetime production. However, people in the West taught it to the Japanese, and W.E. Deming in particular made a big impact in Japan in the 1950's. Japanese industry applied SPC widely and proved that SPC saves money and attracts customers, and industries in the West have been forced to introduce it in order to compete with the Japanese.

SPC is much more than charting. It involves measurement, experimentation, and an objective, data-orientated approach to the study and control of a process. Experimental design plays a key role in SPC, especially as it is applied in the process industries. It will be clear in the later discussion that experimental design is frequently needed to build up single empirical models of processes, and to study the possible causes of variation.

Frontiers in Statistical Quality Control 4
Ed. by Lenz et al.
© Physica-Verlag Heidelberg 1992

SPC cannot be thought of just in statistical terms. The technical side of SPC is clearly essential in gathering information about processes, setting up control charts, and showing how quality or yield can be improved and then maintained at a high level. But as the Japanese have shown, and as practitioners such as Crosby and Deming have also realised, there is a great deal more to it than technical expertise.

3. SHEWHART CHARTS AND THEIR VARIANTS

A basic assumption of the Shewhart chart is that when the process is in control, the data are I I D with a normal distribution with expectation μ and variance σ^2. An out of control condition can be shown by either a change in process average level or process spread. The procedure is therefore to keep charts of the group means, and of group ranges or group standard deviations.

Cumulative sum charts are now frequently used instead of Shewhart charts. A description of all of these charts, with some theory, is given in Wetherill and Brown (1990). The snub-nosed scheme of Rowlands et al (1982), is a particularly useful contribution, and is readily implemented on a computer.

Once an 'out of control' condition is detected, it is assumed that a team can be set up to locate and solve the problem. Clearly, in some industries this step may not be so easy. Questions then arise as to what variables are monitored and at which stage in the process, to make the 'problem solving' step easier.

A vital question to consider here is the relevance of computing to these procedures. There are many situations where the procedures are carried out manually, or else done on a micro by software which simply mimics the manual operation. Clearly, if a computer is used, much more complex rules can be used, and the presentation of the results simplified for use by process operators. In effect, what is wanted is an expert system, which collects data from the process, and gives indications of where the problems might be.

4. BASIC ASSUMPTIONS VERSUS REALITY

4.1 COMPONENT MANUFACTURING INDUSTRIES vs. PROCESS INDUSTRIES

In component manufacturing industries, where the Shewhart chart has been most used, the basic assumptions of I I D normal data frequently hold, and charts devised by Shewhart work well. The processes are frequently well understood, so that although quality problems can be difficult to solve, the procedure is clear. There are numerous published accounts of practical use of Shewhart charts in this context, for example, see Grant and Leavenworth (1972).

Output from these processes is discrete, easily identifiable 'units', which are often manufactured in high volume, for example, transistors, ball bearings etc. Here, obtaining a sample of product for examination presents little problem. Indeed the issue of how to obtain a 'representative sample' is unlikely to arise.

In addition, the measurement procedures required to assess the quality of a unit or product are very frequently quick and easy to carry out, and also very precise, such as measurements of dimensions by micrometer or power output of a circuit. Thus the important components of variation are those which arise directly from the production process and are not masked by problems of sampling and measurement. In the process industries, sampling and measurement are frequently important factors to overcome.

It is also worth remarking that component manufacturing processes are often easily studied by statistical methods of experimental design. The response of the process to a change of a key variable is usually immediate and the resultant output can be measured rapidly. It is therefore quite common to encounter Taguchi-style process optimisation studies where 50 experiments have been carried out during a working day. Would that this be possible in chemical processing industries!

The situation in the process industries is radically different. A common situation is as follows:

a) About ten product variables.
b) Two to five hundred process variable, including pressures, flow areas, purity of catalysts, etc.
c) Twenty input variables.

The effects and relationships of all these variables is often incompletely understood. Thus when problems occur with the process it is not clear which 'knob to twiddle'. Indeed, the problem of locating the cause of the variation is much more complicated.

A typical process is often multi-staged, with continuous recycle around single unit operations and with a proportion of product recycled around many interconnected stages. This results in a highly complex input - output response relationship. Not only do high volume

chemical reactors have significant time constants, but the recycles
induce important delays such that product response to a process
instability can take many hours to stabilise. This clearly has a major
impact on the ability to carry out experiments, especially when it is
important not to disrupt normal production schedules. Thus it is not
uncommon that a single experiment will take 6-8 hours to conduct and so
10 experiments per week might be ambitious on some plant.
Nevertheless, statistical design of experiments have been undertaken
for many decades, in the chemical industry in particular, but in a
fashion rather more subtle than Taguchi methodology.

In process industries, product is often made in bulk, and may
consist of a mixture of components which combine in a homogenous fluid.
That material may also be manufactured under high pressure and become
two-phased (gas-liquid) at lower pressure. Then, obtaining a reliable
representative sample of in-process material is not easy, and in some
situations even hazardous.

Homogenous process product, whether it be liquid or particulate
solid or powder, is most often stored in large storage tanks or silos.
Thus many hours or days of production becomes incompletely mixed before
being eventually tanked off to a customer. Consequently attempting to
trace an unusual final product analysis to an assignable cause can be
rather difficult.

Measurement of the quality of product also frequently presents
difficulties. For example, the measurement acidity is less precise
than the diameter of a component. There are known situations where the
test method gives rise to variability in product measurement which is
more than fifty percent of the apparent total process variation. In
plastics and rubbers manufacture the product is sold on its performance
(eg. flexure strength or impact resistance) when made up into the form
of a component. Thus, while measurement of the composition of the
material is useful in assessing production consistence, it is the
indirect measurement of physical properties which are of most interest.
This can involve a minor manufacturing procedure in itself, such as
making up test plaques of multiple layers of carbon fibre composite
'glued' and compressed together. Clearly the construction of plaques
for test will give rise to sources of variation, quite apart from the
performance of the subsequent physical test.

As a consequence of all these potential sources of variation and
the complexity of process plant, data which is generated for process
and product control is often not understood in detail. Data is also

often positively auto correlated, ie. 'not in statistical control' by the definition advocated by many companies who are encouraging western businesses to make use of traditional SPC methods. Yet this situation is the daily reality for a lot of process industries and so to embark on continuous improvement and make use of statistical process control principles requires special adaptations of the traditional methods. Much of the development work required here is still to be undertaken.

The pattern of variation exhibited by a process industry is therefore likely to be different to that assumed by Shewhart. Instead, there is usually a considerable between-group component of variation, which may take various forms:

   i)   The between-group component is totally random. This is a bit unlikely, but may be useful as an approximation.

  ii)   The between-group component is autocorrelated, due to recycling of material, or due to other less well understood factors.

 iii)   There may be a cyclic component, due to the process itself, daily temperature variation, the presence of automatic controllers, or even due to process operators and shifts.

  iv)   Trends often occur, due to decay of catalysts, the gradual blocking of certain parts of the process, or due to tool wear.

   v)   Sudden jumps may occur, for example, due to fresh batches of raw material.

In addition to these points, the product variables are often of a rather different nature to those used in component manufacturing industries. They are often less precise, and specifications not easy to set. Indeed, it may be difficult to decide how to measure 'quality' in a manner which is directly relevant to the customer.

4.2 SOME MODELS

It will be helpful in clarifying the issues if we state some simple models for the situations discussed above. We let $X_{i,j}$ be a random variable representing the jth observation on the ith group, for $j=1, 2, \ldots, k$, $i=1, 2, \ldots, n$. We write

$$X_{i,j} = \mu + \alpha_i + \epsilon_{i,j} \qquad (1)$$

where $\epsilon \sim N(0,\sigma^2)$ are independent, representing the underlying within-group variation.

Model 1 $\alpha_i = 0$. This is a purely random process.

Model 2 $\alpha_i \sim N(0,\sigma_\alpha^2)$. This is case (i) above.

Model 3 $\alpha_n = \beta_1 \alpha_{n-1} + \beta_2 \alpha_{n-2} + \nu_n$, where β_1, β_2 are constants, and $\nu_n \sim N(0, \sigma_\alpha^2)$.
           This is case (ii) above. Model 3 is used by Vasilopoulos and Stamboulis (1978).

Model 4 $\alpha_i = \beta_1 + \beta_2(i-i_0) + \nu$, where β_1, β_2, i_0 are constants, and $\nu \sim N(0,\sigma_\alpha^2)$. This is case (iv) above.

Model 5   Here we take α_i to be constant for some period with a distribution such as exponential after which it is reset, perhaps at some random value, so that the reset values of α are $\alpha \sim N(0,\sigma_\alpha^2)$.

4.3 VARIANCES RESULTING FROM MODELS

The models just given are listed simply to clarify the following discussion.

Practical processes probably have a mixture of types, with one type often dominating. For models 1 to 4, the group means still have a Normal distribution, but with a different variance.

Model 1 $V(\bar{X}) = \sigma^2/k$

Model 2 $V(\bar{X}) = \sigma_\alpha^2 + \sigma^2/k$ \hfill (2)

Model 3   (k=3) $V(\bar{X}) = \dfrac{\sigma^2}{3}\left\{ 1 + \dfrac{4\beta_1}{3(1-\beta_2)} + \dfrac{2\beta_2}{3} + \dfrac{2\beta_1{}^2}{3(1-\beta_2)}\right\}$ \hfill (3)

Model 4 $\begin{cases} V(\bar{X}) = \sigma_\nu^2 + \sigma^2/k \\ E(\bar{X}) = \beta_1 + \beta_2(i-i_0) \end{cases}$

General formulae for $V(\bar{X})$ for Model 3 are given by Vasilopoulos and Stamboulis (1978). In all cases the R-chart will estimate σ, which will give the correct scaling for the boundaries only in the case of Model 1. Methods for detecting and correcting for certain types of 'extra variation' are given in BS 24. Control charts have been used frequently in the component manufacturing industries, where a combination of Models (1) and (5) above apply. In process industries, the other cases are much more common. One point that is immediately clear is that if we make the wrong assumptions about the type of process we are trying to control, then either in operating Shewhart charts we may be stopping the process with unnecessary frequency, or else we may be losing power. Model (2) above is relatively easy to cope with, but Model (3) is more troublesome, because of the need to 'fit' some type of autocorrelation or moving average model before we can estimate the variance (3).

4.4 COMPONENTS OF VARIANCE

Another feature which complicates applications of SPC in the process industries is the presence of components of variance. Typically several components are present in measurements from a process:

   i)   Local random variations, typically normal.
   ii) Longer term variation in the mean - 'mean wander' to
        engineers. This is frequently autocorrelated, and good fits
        have been obtained using an AR(2) model.
  iii) Measurement error. This is frequently substantial, and may
        include test and laboratory components.
   iv) Sampling error. Again this can be substantial, due to the
        'mean wander' of the process.

Models for process control which do not allow for these components can give spurious results. In particular, sampling error can be very substantial, and sampling powders is notoriously difficult.

4.5 REASONS FOR PROCESS VARIATION

All industrial processes display variation, and there are many different reasons for it, such as:
   (a) Variational noise. This is the variation we observe between
        product manufactured under the same conditions and
        specifications.
   (b) Causes external to the process, such as environmental
        temperature, humidity, etc.
   (c) Process causes. These are due to the process itself, such as
        build up of waste products, ageing of a catalyst, variation
        of loading of a kiln etc.
   (d) Assignable causes of variation. This variation may be due to
        the quality of batches of raw material, incorrect setting of
        equipment, etc.

The procedure adopted in SPC is to try to separate variation that we ordinarily expect of a process, from that which may be due to special or assignable causes. In the component manufacturing field it is frequently possible to assume that the basic variation is independently and Normally distributed. In the process industries, the complexity of the processes means that one usually has to assume a much more complicated structure to the background variation. This makes it rather more difficult to detect 'special causes'.

5.   CHARTING IN THE PROCESS INDUSTRIES

5.1 DEALING WITH EXTRA VARIABILITY

We saw in the previous section that in the process industries, the pattern of variability is in general much more complex than in component manufacturing industries. Unless due account is taken of this when charting, nonsense results. However, the situation has to be handled with care. We shall assume for the moment that data are grouped, in order to clarify the ideas.

If data are grouped, then the first step is to carry out a between and within groups F-test to test for extra variability. If extra variability is present, we need to study the pattern of variation. A time plot will reveal whether or not variation due to special causes is present which may have been a part contributor to the extra variation. A CuSum plot, scaled by the between group variability, will show if important changes in level occurred, and again a search needs to be made for special causes.

After careful investigation, it is likely that we are still left with rather more variability than can be accounted for by the within group variance. In this case charting methods should be based on the between group variance, after eliminating special causes. To use the within group variability in such a situation leads to charts with a large proportion of action points, and brings the whole system into disrepute.

5.2 ONE-AT-A-TIME DATA

When the critical observation is the pH of a chemical, the ppm of a contaminant in a fluid, etc, there is only one observation at a time. Replicate observations show only the local variation of the values, or the laboratory test error, both of which can be small.

The method recommended in the International Standards for getting estimates of variability is to use the moving range of two values method. The average moving range is converted to an estimate of σ by using the usual table constructed for independent Normally distributed variables.

However, there is no reason to limit the moving range to ranges of two. If moving range estimates of σ are obtained on a basis of k values, then the σ estimates in general increase with k.

The reason for this is that such data are often autocorrelated - good fits have been obtained using an AR(2) process. Using k=2 is (almost) equivalent to scaling charts by the within group variability when data is grouped.

The problem of which charts can be operated with such data is one which could be discussed. Plotting deviations from a fitted model is not usually satisfactory due to the complication involved, stability of the model, and the fact that the <u>average level</u> and <u>spread</u> must be controlled, not deviations from a fitted model. A practicable method for this situation is to calculate the σ estimates for a series of k values, study the causes of variation, and choose a suitable $\hat{\sigma}$ by judgment. Moving average charts or exponentially weighted moving average charts should be used. Clearly, it ought to be possible to devise a better scheme.

5.3 THE ARL OF EWMA CHARTS

Since EWMA charts are of considerable value with one-at-a-time data, some ARL tables would be useful to help in the choice of parameters. Unfortunately, no such tables exist, for the reason explained below.

If we denote the observation x_i, i=1, 2, ..., n, and the starting value $m_0 = u$, then the EWMA is given by

$$m_i = (1-p) m_{i-1} + px_i \quad \text{for some} \quad 0 < p < 1 .$$

It is usual to assume action boundaries at $\pm h$, say, so that the ARL formula is

$$L(u) = 1 + \frac{1}{p} \int_{-h}^{h} L(y) f\left[\frac{y - (1-p)u}{p}\right] dy .$$

Crowder (1987) has obtained solutions of this equation.

Unfortunately, there is a problem with this ARL formula. The use of fixed boundaries assumes that the asymptotic variance has been achieved, whereas it takes some while for this limit to be approached. The result is a large number of false alarms at low run lengths.

In the context of process industries, ARL tables are required which allow for starting-up problems, and which show the effects of autocorrelation.

Very similar remarks apply to moving average charts, and again no tables are available; see Lai (1974) for some theoretical results. It is not even clear what a useful definition of ARL might be in these cases.

5.4 SOME POSSIBLE DEVELOPMENTS

Since AR(1) or AR(2) processes have been found to be a good fit in practical cases, the question arises of how best to monitor such processes. One possibility is to fit a model and plot deviations from it. However, there are two problems. One is that it is necessary to monitor the actual mean as well, since a small deviation from a predicted value may still be non-conforming. The second arises from instability in the models. This type of scheme is studied by Alwan & Roberts (1988), and by Montgomery & Friedman (1989).

One suggestion on moving average charts it to improve them by plotting

$$m_i = ((1-p)m_{i-1} + px_i)^+ \quad \text{where} \quad x^+ = \text{Max}(0,x).$$

This gives considerably better 'worst case' properties than the ordinary EWMA chart; see Yaschin (1987).

5.5 MULTIVARIATE DATA

Some better way needs to be found of handling many variables, than simply keeping individual charts. In one plant, it is important to monitor about 20 product variables, and this is done by CuSums. However, the 'false alarm' rate needs to be adjusted, and there is correlation between the variables. One possibility here would be to monitor principal components, but this is not totally satisfactory since the variables have differing importance and specifications.

6. PROCESS CAPABILITY STUDIES

6.1 INITIAL PROCESS CAPABILITY STUDIES

At the outset of SPC implementation a study needs to be carried out on process variability. This involves a number of steps, some of which are summed up in the general methodology given in §7. Key issues are

1) Examining a flow chart of the process, and listing the input, process and output variables
2) Collecting data from various parts of the process
3) Estimating the components of variance present.

In a complicated process, some selection will have to be made of the many variables which might be usefully studied. A key question is then to examine how well the process is capable of meeting specifications on variables.

6.2 SPECIFICATIONS

Typically specifications are not drawn up scientifically. They can vary from the following:

1) What marketing think they want to be competitive.
2) What customers think they require.
3) What competitors claim they can achieve.
4) What the QC department can measure.
5) Historical limits, the reasons for which have long since been lost.

Unfortunately, once limits are established, they tend to be set in stone. Carefully drawn up guidelines of how to set up and maintain specifications do not seem to exist.

6.3 CAPABILITY INDICES

Capability indices are much used in industry. Once specifications have been set, we can define the capability index C_p as

$$C_p = \text{(Allowable range)/(Actual process range)} \tag{4}$$

The denominator is usually interpreted as the range for 99.7% of the distribution. Therefore for data showing only simple Normal variation we have

$$C_p = \text{(Distance between specification)}/6\sigma \tag{5}$$

When the process variability is much more complex the standard definition (5) has no meaning. For between and within group variances σ_B^2 and σ_W^2, we could use

$$C_p = \text{(Distance between specification)}/6\sqrt{(\sigma_B^2 + \sigma_W^2)} \tag{6}$$

If the process measurements have substantial measurement error, then this component must be subtracted from the observed variation in order to get the estimated components for (6). If the variation has a controlled cyclic component, then this should be subtracted. Including a controlled cyclic component can result in very low C_p values even when there is no out of specification product.

The process performance index, C_{pk}, needs similar treatment. We can define this

$$C_{pk} = \text{Min} \left\{ \begin{array}{c} \dfrac{\text{Distance from mean to upper specification limit}}{\text{Half the process range}} \\[2mm] \dfrac{\text{Distance from mean to lower specification limit}}{\text{Half the process range}} \end{array} \right. \tag{7}$$

Again the 'process range' for the denominator is usually interpreted as the range for 99.7% of the distribution.

We note in passing that the definitions (4) and (7) enable us to calculate indices for non-Normal populations, which the standard definitions do not.

It is unfortunate that capability indices have such a strong hold in industry. In the process industries they can be very misleading, particularly when interpreted by staff who understand little about the presence of several components of variation.

7.   GENERAL METHODOLOGY

   7.1 INTRODUCTION

   The key problems with implementing SPC in the process industries are not problems with charts, but the problems of dealing with so many variables. It is not clear what variables to chart, and it is often necessary to build up empirical models of the process to guide staff in corrective action. We give below a scheme which we have tested and used.

   7.2 A STRATEGY FOR IMPLEMENTING SPC

Stage 1 Process Flow
1.1 Draw a schematic diagram of the flow of the process, and note the stages or phases in the process.
1.2 Study the flow of data from the process. Note where and when this data is stored, communication links, etc.

Stage 2 Determine the Problem
2.1 Collect peoples' opinions about the problem, including the customer.
2.2 Determine the important product variables, whether or not they are measured.
2.3 Collect and analyse data on these variables using moving averages, CuSums, and process capability studies.
2.4 Calculate the costs of non-conforming product.

2.5 Interpret the data using process log books, and by consultation with process engineers and operators.

Stage 3   Explore the Process

3.1 Collect information about the process:
   a)   Known from technical sources and reports.
   b)   Relationships or material believed, sometimes strongly
   c)   Conjectures and opinions.
3.2 Break the process down into modules, if possible, and decide on any extra data necessary to achieve this.
3.3 Collect data available from quality control or other routine operations. Decide on extra data required and collect it.
3.4 Analyse and interpret the data using graphs, CuSum plots, multiple regression, multivariate statistical methods, and expert systems (see below).
3.5 Design and carry out experiments on the plant in order to test and establish empirical or theoretical models.
3.6 Choose whether to introduce on-line SPC, process monitoring, or control by other means.
3.7 Choose the variables on which to operate SPC, the types of chart to use, and decide where to put them.
3.8 Implement SPC. This stage will often involve training, and some sort of 'public relations' exercise with staff.

7.3 DISCUSSION

Some success has been found by using principal components. If principal components are calculated for the modules, it is possible to correlate the components, and locate the modules causing most of the variation.

A multidisciplinary team is necessary to carry out this work, involving engineers and the statistician. Ishikawa diagrams have been used to summarise flow chart information etc. If available, 'hypertext' systems can be used very effectively.

8.   THE ROLE OF EXPERT SYSTEMS

8.1   EXPERIMENTAL DESIGN

Although analysis of available process data by using multiple regression and principal components can produce useful information, it is unlikely that very much will be uncovered by these techniques. It

is necessary to use experimental design on the industrial process. As
this is costly, it is desirable to have the designs well thought out.
This leads to the need for design of experiments expertise to be
<u>available to engineers</u>, and this is ideally done through an expert
system.

There are two phases in which expertise is needed. Firstly in
defining the objectives clearly, choosing treatments, etc. This is the
'consultancy' phase of experimental design. Jeffers (1975) produced a
checklist for this, and Williams (1990) has produced an expert system
version called EXPCHECK.

The second phase is the more detailed working out of the correct
design and experimental layout to use and Williams (1990) has developed
a rule elicitation program called KEIRS, Knowledge Elicitation In Rule
Structures. This is the basis of the work in experimental design, and
it can be used to elicit rules for other statistical expert systems.

Following on this work Jallab (1990) has developed new fast
algorithms for experimental designs. His work covers multi-level
factorial experiments, with fractional replication and confounding,
cyclic designs, and response surfaces methodology. It is possible,
using this system, to avoid certain areas of the factor space, etc.
Methods of analysis have been provided by Kelly (1986), who used the
James-Wilkinson algorithm and robust methods.

The combination of these pieces of work into a single useable
system has not been completed, but they do provide a means for making
experimental design more readily available to quality engineers.
Results compare favourably with other systems.

8.2 BEAGLE

Another type of expert system that we have tested and used - with
considerable success - is a genetic evolutionary algorithm system
called BEAGLE, (Forsyth, (1987)). The object of this system is to
discover rules in the data, and it is particularly useful in
discovering the non-linear rules that we have found to be common in the
systems we studied.

The basic method is as follows:

1) The data is divided into two sets; one to breed rules, and the
    other to provide independent validation.
2) An initial population of rules is developed at random.
3) The best of the rules are used for breeding new rules by
    splitting, mating and mutation.

4)   After evaluation of the rules, the worst are discarded and the
     best used to breed a new generation.

A target has to be given to the system, such as to predict failures of a
particular type.

Some care has to be exercised in selecting the variables for this
algorithm. Based on our experience with this, we have a similar system
under construction, which we hope will be even more successful.

8.3   DATA ANALYSIS

In any real industrial process, enormous quantities of data are
produced, and it is clear that some kind of expert system is desirable
for monitoring. Gray (1990) has been working on such a system, linked
into SAS.

Firstly, a 'front-end' to SAS has been produced which uses
principal component analysis to derive data driven rules to be used as
edit checks. The system leads on to simplified access to exploratory
data analysis techniques. This is now being developed into a
formalised expert system containing a strategy for exploratory
analysis.

8.4   STATISTICAL SOFTWARE

Much statistical software is designed either by computer
scientists who have little knowledge of statistics, or by expert
statisticians for use by statisticians. In industrial applications of
SPC much work is done by engineers, who have little formal statistical
training. It is highly desirable that some software become available
which protects users against pitfalls.

Some years ago ICI commissioned a regression package which
included validation checks, checks of outliers, multi-collinearity,
etc, and which had a simple form of output. The results of this work
are written up in Wetherill et al (1986). Similar software is required
for the whole range of techniques used in SPC.

9.   CONCLUSIONS

1.   Standard methods of SPC do not apply in the process industries
without considerable modification. It is necessary to take into
consideration more complex patterns of variation, and several
components of variance.

2.   The usual definition of process capability can be very misleading
in process industry applications. A rather better summary of process
variability is required.

3.   A key problem is that of untangling a complex multivariate dynamic
system. We have developed a methodology which works, but new
approaches are required.

4.   Expert systems have been found to be of considerable value but are
capable of development.

5.   Very little use has been found for sampling inspection. Where it
has been applied, it is usually the application of an Inernational
Standard as part of a legal contract.

6.   It may be that more use ought to be made of incoming inspection.
Audits of producers systems, etc, are sometimes not reliable enough.

7.   SPC should be seen as the application of measurement, and
objective analysis of data, in the control of industrial systems. This
is very much broader than charting.

10. REFERENCES

[1]   ALWAN, L.C, and ROBERTS, H.V., (1988): Time series modelling for
          statistical process control,
          Journal of Business and Economic Statistics, 6, (1), 87-95.

[2]   BAYLY, R.J., (1990) Personal communication.

[3]   CHAMP, C.W. and WOODALL, W.H., (1987): Exact results for Shewhart
          control charts with supplementary runs rules,
          Technometrics, 29, 393-401.

[4]   CROSBY, P, (1985): Quality is free.
          New York, McGraw-Hill.

[5]   CROWDER, S.V., (1987): A simple method for studying run-length
          distributions of exponentially weighted moving average
          charts,
          Technometrics, 29, 401-7.

[6]   FORSYTH, R., (1987): PC BEAGLE User Guide.
  .   Nottingham: Warm Boot Ltd.

[7]   FORSYTH, R. and RADA, R., (1986): Machine learning: Applications
          in expert systems and information retrieval.
          Ellis Horwood,   ISBN 0853 129429

[8]   GRANT, E.L. and LEAVENWORTH, R.S., (1972): Statistical Quality
      Control, 4th edition.
      Tokyo: Ed. McGraw-Hill.

[9]   GRAY, N. (1990): Personal communication.

[10] JALLAB, A.K., (1990): Personal communication.

[11] JEFFERS, J.N.R., (1978): Statistical Checklist: Design of
      Experiments.
      Inst. of Terrestrial Ecology.

[12] KELLY, P.J., (1988): A comparison of classical and robust methods
      of parameter estimation. Thesis.

[13] LAI, T.L., (1974): Control charts based on weighted sums,
      Ann. Statist. 2, 134-47.

[14] MONTGOMERY, D.C., and FRIEDMAN, D.J. (1989): Statistical process
      control in a computer-integrated environment. In Statistical
      Process Control in Automated Manfacturing.
      Eds. Keats, J.B. and Jubele, N.F., New York, Marcel Dekker.

[15] ROWLANDS, R.J. et al (1982): Snub-nosed V-mask control schemes,
      The Statistician, 31, 1-10

[16] VASILOPOLOUS, A.V. and STAMBOULIS, A.P. (1978): Modification of
      control charts limits in the presence of data correlation,
      J. Qual. Tech., 10, No. 1, 20-9.

[17] WETHERILL et al, (1986): Regression analysis with applications.
      London, Chapman and Hall.

[18] WETHERILL, G.B. and BROWN, D.W., (1990) Statistical Process
      Control - Theory and Practice.
      London: Chapman and Hall.

[19] WILLIAMS, M.K., (1987): ExpCheck: an intelligent diagnostic
      system in APL.
      Vector, 4, No. 2, 83-8.

[20] WILLIAMS, M.K., (1988a): An intelligent system for experimental
      design - some ideas,
      Jrnl. Appl. Stats. 15, 325-33.

[21] WILLIAMS, M.K., (1988b): ExpCheck Users Manual.
      University of Newcastle upon Tyne.

[22] WILLIAMS, M.K., (1989): Personal construct analysis and rule
      formation in APL,
      Vector, 6, No. 1, 73-87.

[23] WILLIAMS, M.K. (1990b): On Consultancy and Statistical Expert
      Systems.
      Ph.D. Thesis, in preparation.

[24] YASCHIN, E., (1987): Some aspects of the theory of statistical
      control schemes,
      IBM J. Res & Dev. 31, 199-205.

Part 3
Experimental Design

Comparison of New Techniques to Identify Significant Effects in Unreplicated Factorial Designs

C. Benski, Grenoble, France

1. INTRODUCTION

Industrial experimenters use unreplicated fractional factorial designs whenever, for economical or technical reasons, it is impossible to obtain more than one response for each configuration of the design factors. These designs have usually shared the problem of proper identification of significant effects. This is due to the lack of an independent noise estimate since the variance of the response measure cannot be assessed with just one data point. Also, the use of the multiple interaction estimates to assess the noise is often hindered by the fact that these interactions are, in fractional factorials; confounded with some of the single factors or double interactions one wishes to evaluate. Since an effect is called significant whenever its value is rejected as coming from the same distribution as the noise, the problem is not a trivial one. The classical approach to this problem, first suggested by Daniel in 1959 [1], is based on the usual assumption that all non significant effects are samples of the same normal noise distribution. A further assumpion, the sparsity hypothesis, is that only a small, but unknown, fraction of the computed effects is actually significant. Thus a normal plot of the computed effects should exhibit a straight line behavior except for the significant effects which will appear deviant.

A more recent approach by Box and Meyer [2] is also based on these assumptions but provides a numerical procedure, using Bayes theorem, to compute the a posteriori probability that an effect is significant given its prior probability. This method usually requires intensive numerical integration [3]. Both of these approaches, the normal plots and the Bayesian technique, require subjective judgements from the experimenter: the normal plot, in drawing the best eyeball fit and also in assessing when a point is far enough from the fitted straight line to be considered significant. The Bayesian approach is inherently subjective, by the prior probability that must be specified. For the normal plots, the effect of subjectivity has been shown to be important in terms of both, the straight line eyeball fit [4] as well as for the outlier detection procedure, as shown by Collet and Lewis in Ref. [5]. The relevance of these results to our problem is that, for example, in eye fitting a series of 20 data points, the effficiency can be as low as 32%; the median efficiency being 63%. Notice that most normal plots issued from a fractional factorial design would have about 20 data points (15 and 31 are the most typical). On the other hand the subjective outlier rejection procedure can be better or worse depending on the form of presentation of the data. As for the Bayesian approach of Box and Meyer [2], these authors have shown that, in some cases, the prior probability can modify the conclusions of the analysis. For instance, if the analyst is convinced that a factor has (or does not) have an effect, there is no amount of data that can make him change his mind, (although this is characteristic of all Bayesian approaches.) It is apparent that, in this case, the concept of efficiency is very hard to grasp. In addition, the Bayesian technique is, computationwise, a complex one.

The search for a simpler procedure and the considerations above have recently prompted Benski [6] and Lenth [7] to suggest that real effects could be easily identified, without any subjective judgement, using slightly

Frontiers in Statistical Quality Control 4
Ed. by Lenz et al.
© Physica-Verlag Heidelberg 1992

different forms of formal outlier detection procedures. Benski combined this with a normality test by using Olsson's algorithm [8] to implement Shapiro and Wilk's test of normality [9]. However, the power and efficiency characteristics of the new procedures are unknown. The purpose of this article is to address this aspect of these new approaches.

When applying any statistical test, such as these, the user wishes to be aware of the probability that the test will wrongly reveal the presence of one or more significant effects, given that there are really none as well as the probability that the test will fail to point out to one or more values which are in fact real, i.e. they do not belong to the noise distribution. Notice, however, that there are several alternatives that must be considered. If there are several real effects, the test might identify none, some or all of them without any non real ones. It may also identify a mixture of some real effects and some non real ones. The ideal case would be a test which would identify all the real effects and only these. This is so because a test which identifies all the real effects, simply because it is too sensitive, is of little value to the experimenter. Finally, a test which would correctly identify most of the real effects and only a small percentage of non real ones might still be adequate. This shows that a power analysis must investigate a fairly large number of possible outcomes in order to be useful. The answer to the power problem is mathematically untractable but a Monte Carlo simulation is particularly suitable here and was applied to both techniques.

In Section 2 we present a summary and background for the new proposed procedures: one by this author and the alternative approach proposed by Lenth. The Power Analysis results are presented in Section 3, a comparative application to Taguchi methods is given in Section 4 and our Conclusions are presented in Section 5.

2. NEW EFFECT DETECTION PROCEDURES

It is well known [10] that a fractional factorial design with 2^{n-p} different factor configurations for n factors, each at two levels, will produce, in general using Yates algorithm, $(2^{n-p}-1)$ computed effects. If there is only one measurement for each factor configuration there will not be an independent estimate of the noise since no estimate of the variance of the response will be available. The classical analysis of the variance to obtain the significant effects is obviously impossible.

The procedure suggested by Benski [6] to declare signigicant a computed effect in these cases is as follows. A test of normality on these effects is performed using the algorithm by Olsson [8] which gives a slightly modified version of the Shapiro and Wilk test [9]. If the conclusion of this test is to reject the hypothesis that all the computed effects belong to a normal distribution, an outlier detection procedure, the *fourth-spread* test [11], is used to obtain a combined significance level, using Fisher's combination procedure [12], and assigned to the effect with the largest absolute value. This effect is considered significant and removed from the list of computed effects. The procedure is reiterated with the remaining effects until no rejection is obtained with the normality test. The removed effects are the significant ones. It will be shown that this procedure has good power and efficiency characteristics to identify significant effects. However, the significance levels obtained by Fisher's combination indicated by Benski's procedure are incorrect. This is due to the fact that the test of normality concerns *all* the data and thus the significance level is not applicable to a single point. Nevertheless, it is still correct to select a fixed significance level, say 10%, and perform a formal test of normality. If this test rejects normality at this significance level, the outlier detection procedure can be implemented to identify the significant effects. This approach diminishes the probability of falsely identifying an effect as significant.

There are, however, some restrictions in applying this technique. Since Olsson's algorithm is applicable to samples sizes between 10 and 50, this procedure can only be considered for factorial designs where 2^{n-p} is either 16 or 32, which is a limitation. It should also be noticed that a fractional factorial design with 16 configurations but more than 6 significant effects would also suffer from this limitation since only the largest six effects can be found by this procedure. In these cases, however, the sparsity hypothesis is also no longer tenable. Thus, the main obstacle is really the upper bound. Apart from these limitations, the principle is still valid and any extension of Olsson's algorithm for larger samples could allow the procedure to be applicable to larger designs. The usefulness of this technique stems from the fact that many fractional factorial designs are of sizes equal or smaller than 32. This is so because, even with as many as eleven factors, a design of size 32 is of resolution IV (see e.g., [10]). This means that no main effects will be confounded with two factor interactions, a desirable feature.

The power of the original Shapiro-Wilk test of normality, also known as W Test, has been studied for many non normal distributions [13]. There is, however, little indication of its behavior with respect to the presence of variance shifted values. For example, Pearson, D'Agostino and Bowman [14] have studied the power of the Shapiro-Wilk test for sample sizes 20 and 50 mixed with several proportions of scale contaminated values. But the problem with this study is that it is based on Monte Carlo simulations having only 200 trials. Also, the intensities of the scale shifts considered, from 3 to 7 times the value of the noise distribution, do not necessarily correspond to the typical values found in experimental design experiments. Box and Meyer [2] have shown that the average scale shift in several studies was closer to 10. Be that as it may, the performance of Olsson's algorithm to correctly reject the assumption of normality when the noise distribution is contaminated by one or more outliers remains yet to be fully studied.

Issued from Exploratory Data Analysis [15], the outlier detection technique adopted in Benski's procedure is described by Hoaglin [11]. It is based on the use of the fourth-spread, described next.

Let $E_{(1)}$, $E_{(2)}$, ...,$E_{(n)}$ be the numerical values of the n computed effects sorted in ascending order. The lower and upper fourths, F_L and F_U, are the values of the effects that correspond to the ranks f and $(n+1-f)$ where $f=\frac{1}{2}\lfloor\frac{(n+3)}{2}\rfloor$. The symbol $\lfloor\cdot\rfloor$ stands for the floor operator, i.e. the greatest integer not exceeding the argument. Notice that the values of F_L and F_U must be obtained by interpolation whenever the sample size n is congruent to 0 or 3 (modulo 4). The fourth-spread d_F is the difference between F_U and F_L. The outlier detection procedure is to declare as outlier any value outside of the interval $[F_L-1.5d_F \ ; \ F_U+1.5d_F]$. Now, the basic hypothesis of factorial designs is that effects which are not significant are samples from a normally distributed noise function, $N(0,\sigma)$, with a mean equal to zero and unknown variance σ^2. Therefore, the significant effects will be contaminant values issued from another normal distribution, $N(0,k\sigma)$, i.e., from a variance or scale shifted distribution. Within this context, the criterion for identifying a significant effect is that its value should lie outside of the interval $[-2d_F \ ; \ +2d_F]$.

Hoaglin, Iglewicz and Tukey [16] have studied this outlier detection procedure for some sample sizes between 5 and 50 by a Monte Carlo approach. They have obtained the fraction of values that lie, on the average and for a single sample, outside of the above interval. This was approximated by a conservative formula, given as a function of sample size, which was then used by Benski [6] for the Fisher combination procedure mentioned before. They have also obtained the proportion of samples that do not contain any values outside of the interval. Iglewicz [17] has given results describing the behavior of this technique with respect to a single contaminant value issued from a scale shifted distribution having a value of $k=100$. As with the normality test, this value of the scale shift is not representative of what is usually seen in factorial design experiments. Although all of these results contribute somewhat to our knowledge of the performance of the outlier detection procedure, a more detailed analysis is clearly needed, in particular with respect to the presence of more than one outlier.

Another unknown characteristic of the fourth-spread outlier test is its behavior upon iteration. Now, the basic principle of the fourth-spread is to estimate the standard deviation by giving greater weight to the central values of the sample, presumably the uncontaminated ones. The iteration algorithm recomputes this standard deviation each time an outlier is removed from the sample. The impact of this iteration on the effectiveness of the test to identify the outliers is not known and had to be investigated. Some results concerning this problem are given in the next Section.

The other technique for the identification of significant effects by Lenth [7], is a pure outlier detection approach without the preliminary normality test. It also uses the data closest to the median to compute a pseudo-standard error (PSE) to obtain two types of cut-out points. The PSE is given by:

$$PSE = 1.5 \times \underset{|E_{(j)}| < 2.5 s_0}{median} |E_{(j)}| \tag{1}$$

where $s_0 = 1.5 \times \underset{j}{median} |E_{(j)}|$.

Lenth suggests the following two types of cut-out points: the first inner one is given by the product of PSE times $t_{1-\frac{\alpha}{2}, \nu}$, where t is the fractile of Student's distribution function with $\nu = \frac{n}{3}$ degrees of freedom. This would give an α significance level for effects which are greater in absolute value than this product. A second outer cut-out point is also given to take into account the fact that one is making a *simultaneous* statement about all the effects found to be significant. These cut-out points, at the 95% confidence level, are given by the product of the PSE times the γ fractile of Student's t distribution function also with $\nu = \frac{n}{3}$ degrees of freedom: $t_{\gamma; \nu} \times PSE$, where $\gamma = (1 + .95^{1/n})/2$. All the effects whose absolute values are greater than $t_{\gamma; \nu} \times PSE$ are considered significant at the specified confidence level. In the power analysis we have used these outer cut-out points.

3. POWER ANALYSIS

We have concentrated our study on mainly two sample sizes: 15 and 31. This is so because of the present limitations of Olsson's algorithm to sample sizes less than 50. Hence, the sizes one could analyze from a result of a 2^{n-p} factorial design experiment would be either 15 or 31. In Lenth's procedure this restriction is not necessary. For the number of outliers j, we have investigated up to three contaminants. We have thus generated $(n-j)$ normal deviates from a $N(0,1)$ distribution mixed with j normal deviates from a $N(0,10)$ distribution with n equal to 15 and 31 and j equal to 1, 2 and 3. The value $k=10$ for the scale shift was adopted because, as mentioned before, it is typical of many factorial experiments.

Table 1 shows, as a function of the real effects generated j, the number of times each test rejected correctly the hypothesis that each sample came from a pure normal distribution. This table was obtained using 10 000 Monte Carlo trials each time. The tests applied were: the W Test alone (Olsson's algorithm was always applied at the 10% significance level) under the column labeled W, the fourth-spread outlier detection test alone, under the column labeled d_F, under the column labeled *combin.*, the number of samples rejected by both, the W Test and by the outlier detection test and finally, under PSE the results of Lenth's procedure using the simultaneous cut-out points. The power of each test is seen to increase, as expected, as a function of the number of contaminant values. The sample size contributes only marginally to the power. This table also shows that if a sample is rejected by the W Test, it is very likely that it will also be rejected by the fourth spread test. Notice, though, that the converse is not true, see Table 2. On Table 1, the alternative to the null hypothesis is that the sample contains, at least, one wild value. This is meant to give an overall idea of the sensitivity of these tests, more than their ability to actually identify the correct outliers. Lenth's procedure is consistently the worst, independently of the number of real effects.

j	$n=15$				$n=31$			
	W	d_F	combin.	PSE	W	d_F	combin.	PSE
1	7355	8270	7275	5932	7451	8412	7394	6818
2	9108	9545	9084	8123	9218	9644	9199	8849
3	9591	9860	9580	9009	9729	9908	9726	9583

Table 1: Rejection counts in Monte Carlo trials of size 10 000 for a scale shift $k=10$ as a function of the number of real effects. The number of times each test rejected the null hypothesis for samples containing $j=1$, 2 and 3 real effects issued from a $N(0,k)$ distribution is given for sample sizes $n=15$ and 31. W corresponds to the W test, d_F to the fourth-spread test, combin. to the number of samples rejected combining the W and the d_F tests and PSE are the samples rejected by Lenth's outer cut-out points.

Table 2 shows the number of times each of these procedures wrongly rejected the null hypothesis when applied to samples issued exclusively from $N(0,1)$. Lenth's procedure is significantly better in giving the smallest type I error. Notice, however, that the type I error obtained by the combined procedure proposed by Benski is essentially that of the normality test which was chosen as 10%.

n	W	d_F	combin.	PSE
15	982	2895	713	220
31	953	3402	743	295

Table 2: Number of Type I errors for each test. 10 000 samples of size 15 and 31, issued from $N(0,1)$, were subjected to the W Test only, the W column, to the fourth-spread Test only, the d_F column. The combin. column shows the number of times both tests rejected wrongly the null hypothesis. PSE is the result of Lenth's test using the outer cut-out points.

The fourth spread outlier detection test gave a rate of false rejections of about 30% for $n=15$ and 34% for $n=31$. This seemed somewhat too high compared to the results of Hoaglin, Iglewicz and Tukey [16] who also used 10 000 trials for their experiments. These authors obtained Type I errors of 23.2% for $n=20$ and 29% for $n=30$.

We therefore conducted an experiment with these sample sizes, (20 and 30), and were able to confirm the accuracy of their results. Thus, the fourth spread test test has a non monotone behavior as a function of sample size. This was also observed by Hoaglin, Iglewicz and Tukey. In spite of this feature, which needs further research, the fourth spread outlier detection technique is able to reduce the Type I error (false rejections) of the W Test alone by, at least, 20%, as shown by the combination column of this table where the number of false rejections drops from 982 to 713 for $n=15$ and from 953 to 743 for $n=31$.

In tables 3, 4 and 5 we have studied the efficiency with which each test not only detects, but also identifies, the presence of eventual outliers, all issued from the distribution $N(0, 10)$. As we mentioned in the section describing the new procedures, the effect of applying a test to identify outliers can result in a mixture of true and false statements. For example, for the $j=3$ case, we have to count the number of times the procedure identifies correctly the three real effects and nothing else, the number of times it identifies the three effects but also, incorrectly, some other effects which actually belong to the noise distribution $N(0,1)$, the number of times it identifies just two of the three effects and nothing else, the number of times it identifies two of the three plus some other non real effects and so on until, finally, we count the number of times the procedure does not identify any effects at all or that none of the ones identified are real. It was possible to summarize these results by cumulating in these tables the number of times each test was exhaustively exact in Table 3, (no spurious effect was identified along with the real ones which were all correctly identified), then, in Table 4, the number of times each test identified only real effects, some or all of them. Finally, in Table 5, we give the number of times each test was totally wrong because it either did not reject the null hypothesis or, if it did, it was for the wrong reasons. These last two tables show that Lenth's procedure was poorer than the combined normality and fourth spread tests to identify real effects.

We have also included two columns for the fourth spread outlier detection test: one labeled $d_{F_{it}}$ and one labeled d_F, to indicate the performance of the iterated and non iterated fourth-spread tests. The reason for this is that the non iterated test seemed to perform fairly well on its own, specially when one considers its simplicity. With this test, a single application of the outlier detection procedure would identify all significant effects without any further calculations! Were it not for its large Type I error, it might even be considered as a stand alone alternative to the W Test when the alternative is the presence of scale contaminated outliers. But then this is where Lenth's test would be a better choice. In the iterated procedure, only the effect with the largest absolute value is removed after each iteration, even if more effects are signaled by the outlier detection test. Its power is worse than the W Test. But the application of the combination procedure imposes an iteration since the W Test, as opposed to the outlier detection procedure, does not single out particular effects as contaminants; it simple indicates the non-normal nature of the sample data. This is the reason why Fisher's combination formula cannot be applied.

Because we noticed, in Table 3, that the iterated and non iterated fourth-spread procedures had such a different behavior as a function of the number of contaminant values, we carried the investigation further to 4 and 5 real outliers, just for the outlier detection procedures. We discovered that the iterated procedure decreased in power with an oscillatory curve of period equal to two, i.e., the result for 2 and 3 outliers was similar, then 4 and 5 came further down. This was true independently of the sample size. This behavior is in sharp contrast with the decrease in power of the non iterated procedure which had a monotone and smooth behavior. We were unable to find a compelling explanation for this difference.

Table 4 shows the ability of the different procedures to give results which, although perhaps not complete, contain no false statement. The performance of the combined procedure was, within the accuracy of the simulation, always equal or better than any of the others, including Lenth's. In particular, the power of the combined procedure is seen to be enhanced with respect to the W Test alone. The (small) price to pay is exhibited in Table 5: the combination procedure increases very slightly the probability that the conclusions be totally wrong, either because no effects have been identified or because the ones identified are all wrong. Further work to compare Benski's procedure to Lenth's, but using a 5% significance level for the normality test, remains to be done. The expected effect of this would be to obtain comparable type I errors but, perhaps, still somewhat better power for the combined test than for Lenth's approach. Further research is also needed to measure the power of Lenth's procedure when combined with the normality test.

j	n	$combin.$	W	$d_{F_{it}}$	d_F	PSE
0	15	9287	9018	7105	7105	9780
	31	9257	9047	6598	6598	9705
1	15	6653	6421	6004	6177	5739
	31	6695	6529	5553	5551	6490
2	15	4743	4637	3904	5067	3529
	31	4808	4726	3874	4602	4390
3	15	3361	3343	3766	3851	2145
	31	3489	3328	3351	3721	2904

Table 3: Number of times that each test correctly identifies all the real effects and only these. That is, only those j effects belonging to $N(0,10)$. W is the result of the normality test alone, $d_{F_{it}}$ is the iterated outlier detection procedure. d_F is the same test but without iteration, $combin.$ is the application of Benski's test and PSE is the application of Lenth's test using the outer cut-out points.

j	n	$combin.$	W	$d_{F_{it}}$	d_F	PSE
2	15	8437	8188	6912	8185	7959
	31	8587	8391	6572	7234	8593
3	15	9024	8785	7811	9072	8872
	31	9139	8906	7080	8050	9158

Table 4: Number of times that each test identifies either all or some of the j real effects and only these. This table is only necessary for $j>1$. Notation as in Table 3.

j	n	combin.	W	$d_{F_{it}}$	d_F	PSE
1	15	2838	2797	2113	2106	4102
	31	2761	2730	2070	2125	3283
2	15	981	962	580	553	1884
	31	800	788	467	526	1159
3	15	443	416	166	166	996
	31	256	232	125	117	418

Table 5: Number of times each test does not identify any of the real effects. Notation as in Table 3.

4. APPLICATION

A typical case of unreplicated experimental designs is found in the implementation of Taguchi methods [18] to reduce process variability. The experiment is designed in these cases to determine which, if any, of the controlled factors can be used to obtain a more uniform response. Although this response is measured several times, the purpose of this replication is just to obtain a single estimate of the variance at a given setting of the controlled factors. The measurement of the variance is therefore non replicated and noise estimation becomes a problem.

The following example is given to illustrate the topic of subjectivity in Taguchi type experiments. Kackar and Shoemaker [19] performed such en experiment in order to reduce the nonuniformity of the epitaxial layer in the manufacturing process of silicon wafers used in the semiconductor industry. Eight controllable factors were involved in the 2_{IV}^{8-4} fractional factorial experiment. The negative of the natural logarithm of the estimated variance was here the unreplicated response. The results of this experiment are shown in standard order, in Table 6. The computed main effects as well as the two factor interaction confoundings are given in ascending order in Table 7. For simplicity, higher order interactions were omitted from this table.

Figure 1 shows the stabilized normal plot [20] corresponding to Table 7. This type of plot is equivalent to a normal plot, except that a unit square is used. An arcsin transformation of the effects and the cumulated fraction i/n, where i is the order of the observed effect, allows for this very simple presentation.

We have purposedly omitted the drawing of a straight line through the data points. Before reading the next paragraph, see how many significant effects, if any at all, you can suggest based on this plot. Show the drawing straight up and up-side down to some of your colleagues and see how many wild dots, i.e. deviating from a straight line, they can spot. We have performed this test with several of our co-workers and found a large spread in terms of the identified effects. Scale and orientation of the plot modified the answers from some of them! Compare now your own conclusions to the possible answers given by the different approaches described below. The meaning of subjectivity should become now apparent.

Run	A	B	C	D	E	F	G	H	$-\log s^2$
1	−	−	−	−	−	−	−	−	0.4425
2	−	−	−	−	+	+	+	+	1.1989
3	−	−	+	+	−	−	+	+	1.4307
4	−	−	+	+	+	+	−	−	0.6505
5	−	+	−	+	−	+	−	+	1.4230
6	−	+	−	+	+	−	+	−	0.4969
7	−	+	+	−	−	+	+	−	0.3267
8	−	+	+	−	+	−	−	+	0.6270
9	+	−	−	+	−	+	+	−	0.3467
10	+	−	−	+	+	−	−	+	0.8563
11	+	−	+	−	−	+	−	+	0.4369
12	+	−	+	−	+	−	+	−	0.3131
13	+	+	−	−	−	−	+	+	0.6154
14	+	+	−	−	+	+	−	−	0.2292
15	+	+	+	+	−	−	−	−	0.1190
16	+	+	+	+	+	+	+	+	0.8625

Table 6: Results of the epitaxial layer experiment. The reponse is the negative of the natural logarithm of the variance. The generators for this design were $D=ABC$, $F=ABE$, $G=ACE$ and $H=BCE$.

Factor	Computed effect
H. Nozzle position	−0.5658
D. Deposition time	−0.2495
$AE+BF+CG+DH$	−0.1741
G. HCl flow rate	−0.1008
$AB+CD+EF+GH$	−0.0903
F. HCl etch temperature	−0.0717
$AC+BD+EG+FH$	−0.0263
$AG+BH+CE+DF$	−0.0233
E. Arsenic gas flow rate	−0.0117
$AF+BE+CH+DG$	+0.0788
$AD+BC+EH+FG$	+0.1020
C. Deposition temperature	+0.1053
B. Code of wafers	+0.1220
$AH+BG+CF+DE$	+0.1250
A. Susceptor rotation method	+0.3521

Table 7: Computed effects of the epitaxial layer experiment sorted in ascending order. Only single factors and double interactions are listed.

Epitaxial layer experiment

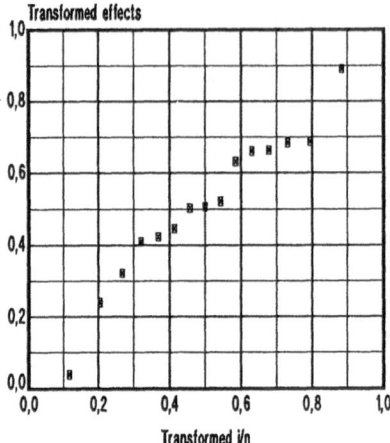

Figure 1: Stabilized normal plot of the computed effects in the epitaxial layer experiment. Plotted points are in the same ascending order as in Table 7, from lower left to upper right, (see [20]).

In their paper the authors used the relative magnitude of each effect as a subjective but simple outlier detection rule to identify significant effects. No normal plot is presented in this paper. They chose the two largest effects, (nozzle position and susceptor rotation method), as being tentatively important and conducted a confirmation experiment: all factors were left in their initial settings, except the chosen two, which were modified in the sense suggested by the experimental results. The variance was reduced by 61% with the new settings.

For comparison purposes we performed first a Bayesian analysis on the computed effects. The prior probability, α, for each effect is set at $\alpha=0.2$, a common default value. After five minutes of heavy computation using an AT compatible machine, the posterior probabilities showed that the only effect that could be considered significant was the nozzle position. This factor reaches a posterior probability of 94.3% while the next posterior probability, corresponding to the susceptor rotation method, is 64%. All the others were lower than 50%. To assess the effect of subjectivity we increased the prior probability of the susceptor rotation factor to $\alpha=0.4$. The posterior probability of this factor, 92%, becomes comparable to the nozzle factor at $\alpha=0.2$ and both effects can now be considered significant.

The fifteen computed effects were then put through Benski's combined identification procedure. The normality of the data could actually be rejected at the 8.9% significance level.. Once the largest effect in absolute value is removed, the nozzle position, the normality test cannot reject the null hypothesis at the 10% significance level. Thus, no other remaining factor or interaction is considered significant. In addition, assuming a model where only this factor is used to predict the response, the normality analysis of the residuals showed that this model could not be rejected, even when a large significance level was adopted.

The fourth-spread outlier detection procedure gives $2d_F=0.4$. Thus, the only significant effect identified by this procedure, (larger in absolute value than 0.4), corresponds to the same factor given by the normality test, i.e., the nozzle position. No more iterations are needed since only one outlier value was found. The significance level here can be conservatively approximated by: $P = 0.00698 + \frac{0.4}{15} = 0.03365$, [11].

Our conclusion using the combined procedure is that a change in nozzle position is enough to reduce the variability of the response.

Using Lenth's procedure we obtain the *simultaneous* cut-out points at ± 0.79. Had an experimenter used these outer cut-out points the conclusion would have been that no effects are significant since they are all smaller than this value. For the inner cut-out points we obtain ± 0.39 and then the nozzle position appears as the only factor that can be considered significant. Since there are no simultaneous statements the conclusion would be to retain just this factor. However, this looks (and is) a post-hoc type of analysis.

The conclusions reached by the different methods of analysis are summarized in Table 8.

Effect	Subjective Outlier	Bayesian ($\alpha=0.2$)	Bayesian ($\alpha=0.4$)	*combin.* $W+$ fourth-spread	PSE proc.
H	*yes*	94.3%	99.9%	>97%	*no*
A	*yes*	*no*	92%	*no*	*no*
the rest	*no*	*no*	*no*	*no*	*no*

Table 8: Confidence level at which effects are identified as significant according to different techniques. Subjective outlier is the technique used in the article by Kackar and Shoemaker [19]. Normal plot results were omitted because of ambiguous opinions.

It is, of course, impossible to say now how many effects are "really" present and thus determine which technique is "right". Engineering judgement, previous information or subjective bias might lead the experimenters to a conclusion which is not necessarily supported by evidence from the data. Although we have a preference for non subjective techniques, there are times when there is some interest in the knowledgeable application of subjective ones. For example, the normal plots have been shown to be useful not only for the identification of effects but also to identify systematic errors, [21]. In some cases the Bayesian approach could also have its merit. There is no reason to confine the analysis to just one technique. However, this application illustrates well the point we wished to make: either one of the new procedures can be a valuable tool to reduce the influence of subjectivity in Taguchi type experiments where noise estimates are often unavailable. The ones based on outlier detection techniques have the advantage of being computationally simple when compared to the Bayesian technique.

5. CONCLUSIONS

The power characteristics of two new procedures to identify significant effects in unreplicated fractional factorial experiments have been studied. The normality test combined with the fourth spread outlier detection and the PSE test have been shown to be a valuable complement to the classical but subjective normal plots. They are also much simpler, in terms of computational effort, than the Bayesian technique and do not need a subjective prior probability to assess significance. The technique based on the PSE, although less powerful, is simpler than all the others. Further research is needed to assess the power of these techniques as a function of the scale shift of the contaminant values.

6. REFERENCES

[1] DANIEL, C. (1959), "Use of Half-Normal Plots in Interpreting Factorial Two-Level Experiments," *Technometrics*, 1, 311-341.

[2] BOX, G. E. P., and MEYER, R. D. (1986), "An Analysis for Unreplicated Fractional Factorials," *Technometrics*, 28, 11-18.

[3] STEPHENSON, W. R., HULTING, F. L., and MOORE, K. (1989), "Posterior Probabilities for Identifying Active Effects in Unreplicated Experiments," *Journal of Quality Technology*, 21, 202-212.

[4] MOSTELLER, F., SIEGEL, A. F., TRAPIDO, E., and YOUTZ, C. (1983), "Fitting Straight Lines by Eye." In *Understanding Robust and Exploratory Data Analysis*. (D. C. Hoaglin, F. Mosteller and J. F. Tukey, Eds.), John Wiley & Sons, New York, NY.

[5] COLLET, D., and LEWIS, T. (1976), "The Subjective Nature of Outlier Rejection Procedures," *Applied Statistics*, 25, 228-237.

[6] BENSKI, H. C. (1989), "Use of a Normality Test to Identify Significant Effects in Factorial Designs," *Journal of Quality Technology*, 21, 174-178.

[7] LENTH, R. V. (1989), "Quick and Easy Analysis of Unreplicated Factorials," *Technometrics*, 31, 469-473.

[8] OLSSON, D. M. (1979), "A Small-Sample Test for Non-Normality," *Journal of Quality Technology*, 14, 95-99.

[9] SHAPIRO, S. S., and WILK, M. B. (1965), "An Analysis of Variance Test for Normality (Complete Samples)," *Biometrika*, 52, 591-611.

[10] BOX, G. E. P., HUNTER, W. G., and HUNTER, J. S. (1978), *Statistics for Experimenters*, John Wiley & Sons, New York, NY.

[11] HOAGLIN, D. C. (1983), "Letter Values: a Set of Selected Order Statistics." In *Understanding Robust and Exploratory Data Analysis*. (D. C. Hoaglin, F. Mosteller and J. F. Tukey, Eds.), John Wiley & Sons, New York, NY.

[12] STEPHENS, M. A. (1986), "Tests for the Uniform Distribution." In *Goodness-of-Fit Techniques* (R. B. D'Agostino and M. A. Stephens, Eds.), Vol. 68, Marcel Dekker Statistics Textbooks and Monographs, New York, NY.

[13] SHAPIRO, S. S. (1980) *How to Test Normality and Other Distributional Assumptions*, Vol. 3, ASQC Basic References in Quality Control: Statistical Techniques, American Society for Quality Control, Milwaukee, WI.

[14] PEARSON, E. S., D'AGOSTINO, R. B., and BOWMAN, K. O. (1977), "Tests for Departure from Normality: Comparison of Powers," *Biometrika*, 64, 231-246.

[15] TUKEY, J. W., (1977), *Exploratory Data Analysis*, Addison-Wesley Publishing Company, Reading, MA.

[16] HOAGLIN, D. C., IGLEWICZ, B., and TUKEY, J. W. (1981). "Small-sample Performance of a Resistance Rule for Outlier Detection," 1980 *Proceedings of the Statistical Computing Section*. American Statistical Association, Washington, D.C., 148-152.

[17] IGLEWICZ,B. (1983), "Robust Scale Estimators and Confidence Intervals for Location." In *Understanding Robust and Exploratory Data Analysis.* (D. C. Hoaglin, F. Mosteller and J. F. Tukey, Eds.), John Wiley & Sons, New York, NY.

[18] TAGUCHI, G., (1986), *Introduction to Quality Engineering*, Asian Productivity Organization, Tokyo.

[19] KACKAR, R. N., and SHOEMAKER, A. C. (1986), "Robust Design: A Cost-Effective Method for Improving Manufacturing Processes," *AT&T Technical Journal*, 65, 39-50.

[20] NELSON, L. S. (1989), "A Stabilized Normal Probability Plotting Technique," *Journal of Quality Technology*, 22, 213-215.

[21] BOX, G. E. P., and MEYER, R. D. (1987), "Analysis of Unreplicated Factorials Allowing for Possibly Faulty Observations." In *Design, Data & Analysis* (C. L. Mallows Ed.), John Wiley & Sons, New York, NY.

An Algorithm for the Construction of Orthogonal Factorial Designs

M. Riebschläger and P.-Th. Wilrich, Berlin, Germany

1. Introduction

Experiments are often performed in order to examine the dependence of a response variable Y on one or several factors described by the linear model

$$Y = X\beta + u$$

with

$$
\begin{array}{lll}
Y & \in \mathbf{R}^N & \text{observation vector,} \\
N & \in \mathbf{N} & \text{number of observations,} \\
\beta & \in \mathbf{R}^p & \text{vector of unknown parameters,} \\
p & \in \mathbf{N} & \text{number of unknown parameters,} \\
X & \in \mathbf{R}^{N \times p} & \text{design matrix,} \\
u & \in \mathbf{R}^N & \text{vector of residuals, with the assumptions}
\end{array}
$$

$E(u) = 0$ and $\mathrm{Cov}(u) = \sigma^2 I_N$.
($\sigma^2 \in \mathbf{R}^+; I_N : N \times N$ identity matrix).

Let $\mathcal{M} := \mathcal{L}\{X\}$ be the linear space spanned by the column vectors $x_1, \ldots, x_p$ of the design matrix X (the model-space) and let $\mu := E(Y) = X\beta \in \mathcal{M}$ be the expectation of Y. The OLS estimate of μ is obtained by the orthogonal projection of Y onto $\mathcal{M}$, i. e. $\hat{\mu} = P_{\mathcal{M}} Y$.

In order to investigate the influence of a factor, the variable must be examined on at least two levels. If several factors are involved and experiments are performed for all factor level combinations, all main effects and interactions can be estimated. However, if the number of factors involved in the experiment is large, the number of factor level combinations is also large and hence the experimental effort is voluminous. Very often, by prior knowledge, several effects, especially higher-order interactions, can be neglected. In those cases it may be sufficient to assign only a part of all possible factor level combinations to experimental units or — in case of a block design — to assign only a fraction to the blocks.

The algorithm, described in the following sections, generates orthogonal fractional and block designs, which supply uncorrelated estimates for specified effects to be included in the model for both analysis of variance and regression models. There is no restriction to the numbers of levels of the factors denoted by $p_1, \ldots, p_K$ (where K is the number of factors), and hence the algorithm is a generalization of the following ones :

1) Greenfield (1976) : $p_1, \ldots, p_K = 2$
2) Franklin 1985) / Turiel (1988) : $p_1, \ldots, p_K = p$
   $$p = 2, 3, 5, 7$$
3) Rasch (1986) : $CADEMO$
   $$p_1, \ldots, p_K = p, \; p \text{ prime}$$

Frontiers in Statistical Quality Control 4
Ed. by Lenz et al.
© Physica-Verlag Heidelberg 1992

4) Patterson (1976) : *DSIGN*
all combinations of $p_1, \ldots, p_K = p$ are
admitted but the defining contrast
(see section 3) has to be specified by the user

5) Zemroch (1986) : *SELINA*
resolution-r-designs (all effects up to order
$r - 2$ are orthogonal (uncorrelated))

2. Two-level main-effect models

For models with two levels per factor the method for the construction of the designs is based on Hadamard-matrices. These designs are well known as Plackett-Burman-designs and supply uncorrelated main-effects estimates (Plackett, Burman, 1946).

A square matrix H_n of order n whose entries are $+1$ or -1 is called a Hadamard matrix of order n, provided that its columns are pairwise orthogonal, in other words $H'H = nI$.

It has been conjectured, that a Hadamard matrix exists for every $n = 4j, j \in \mathbf{N}$, but no general proof is available. The smallest order, which is undecided, is 268, but this is far beyond the range of relevance concerning experimental design theory.

For models including K factors, the minimum sample size of a Plackett-Burman-design is $n \geq K + 1, n = 4j, j \in \mathbf{N}$.

The corresponding Hadamard matrix H_n has to be transformed such that the first column is the vector $\mathbf{1}$, having all its entries $+1$. The remaining columns represent the design, and the factor level combinations are given by the rows.

Whenever $n > K + 1$, K columns have to be selected at random (in addition to the first one).

Example :

For $K = 10$ a Plackett-Burman design is given by the Hadamard matrix H_{12} in which one of the columns $2, 3, \ldots, 12$ chosen randomly has to be deleted.

$$
H_{12} = \begin{bmatrix}
+ & + & + & + & + & + & + & + & + & + & + & + \\
+ & - & + & - & + & + & + & - & - & - & + & - \\
+ & - & - & + & - & + & + & + & - & - & - & + \\
+ & + & - & - & + & - & + & + & + & - & - & - \\
+ & - & + & - & - & + & - & + & + & + & - & - \\
+ & - & - & + & - & - & + & - & + & + & + & - \\
+ & - & - & - & + & - & - & + & - & + & + & + \\
+ & + & - & - & - & + & - & - & + & - & + & + \\
+ & + & + & - & - & - & + & - & - & + & - & + \\
+ & + & + & + & - & - & - & + & - & - & + & - \\
+ & - & + & + & + & - & - & - & + & - & - & + \\
+ & + & - & + & + & + & - & - & - & + & - & -
\end{bmatrix}
$$

The first row gives the factor level combination where all factors are on their high level, and if for instance the last column has been deleted, the second row indicates, that for the next experimental unit the factors 1, 3, 7, 8, 9 are at their low level and the others at their high level, etc.

In addition to the orthogonality of the main effect estimators, these will not be biased by two-factor interactions in a so called fold-over design consisting of two Plackett-Burman-designs, where the factor level combinations of the first Plackett-Burman-design are replicated with reversed sign.

3. Factorial designs — the general case

For models with more than two levels per factor, or models including interactions, a different method of construction has to be used.

Those interactions, that are to be included in the model in addition to the main effects, have to be be specified. The total of these interactions and the main effects are called "effects of interest". The remaining effects are called "negligible effects".

The method for the construction of orthogonal fractional and block designs is based on aliasing effects of interest and negligible effects in a systematic way described below.

Notation:

For a K-factor model with number of levels $p_1, \ldots, p_K$, the combinations of factor levels are represented as vectors $x = (x_1, \ldots, x_K)$, where $x_i \in \{0, \ldots, p_i - 1\}$ indicates the level of factor i. $D = \{x; x = (x_1, \ldots, x_K), x_i = 0, \ldots, p_i - 1, i = 1, \ldots, K\}$ is the set of all possible combinations of factor levels.

Effects are also represented as vectors $z = (z_1, \ldots, z_K)$, where the nonzero coefficients of z indicate the factors involved in the effect.

Let Π be the set of all vectors z and Π^* the set of all vectors z belonging to effects of interest.

Method :

Let γ be the least common multiple of the numbers of factor levels $\gamma := lcm[p_1, \ldots, p_K]$. Then the scalar product defined by

$$[z, x] := \sum_{j=1}^{K} x_j z_j \frac{\gamma}{p_j} \bmod \gamma, \text{for a fixed } z \in \Pi$$

generates a partition $P(z) = \{D_1, \ldots, D_{n(z)}\}$ on D for every vector z (Bailey, Gilchrist, Patterson, 1977).

By means of this partition the model space $\mathcal{M} = \mathcal{L}\{X\}$ can be divided into orthogonal subspaces each of them belonging to the effect represented by the vector z used to generate the partition. The estimates of the effects are obtained by orthogonal projections onto these subspaces. In a block design the blocks correspond to classes, whereas in a fractional design only one class D^* is selected. The vector z^* chosen in order to generate the partition is called defining contrast.

For fractional designs the model space $\mathcal{M}$ is restricted on the class D^*, and the subspaces are restricted in an analogous way. Due to the restriction some subspaces coincide, whereas the others remain orthogonal, and hence the estimates of the corresponding effects remain uncorrelated.

For block designs, the effects belonging to z^* are confounded with block effects.

In order to keep subspaces belonging to effects of interest orthogonal, the defining contrast z^* has to fulfill the following properties:

1. $\lambda z^* \notin \Pi^*$

2. $\lambda' z' \neq \lambda'' z'' + \lambda z^*, \text{f. a. } z', z'' \in \Pi^*$

For block designs only condition 1 must hold.

By selecting additional defining contrasts and building subsections of classes of the corresponding partitions, a further reduction of the sample size can be achieved, if vectors $z_1^*, \ldots, z_n^*$ can be found that fulfill the property

$$\lambda' z' \neq \lambda'' z'' + \sum_{i=1}^{n} \lambda_i z_i^* \text{ f. a. } z', z'' \in \Pi^*$$

where multiplication and addition of the z-vectors is defined by

$$\lambda' z' + \lambda'' z'' := \begin{pmatrix} \lambda' z'_1 + \lambda z''_1 \bmod p_1 \\ \vdots \\ \lambda' z''_K + \lambda'' z''_K \bmod p_K \end{pmatrix}.$$

The number of elegible defining contrasts depends on the constellation of the numbers of factor levels $p_1, \ldots, p_K$. No reduction of sample size can be achieved, when the numbers of factor levels have no common divisor, whereas best results are obtained for models with identical numbers of levels $p_1 = \ldots = p_K = p$ with p a prime number.

4. The algorithm

The algorithm which is described in detail in Riebschläger (1989), requires the following input :

1. K (number of factors)

2. $p_1, \ldots, p_K$ (numbers of factor levels)

3. interactions to be included into the model

4. block or fractional design

5. sample size

If the sample size, requested by the user, cannot be achieved, a feasible sample size (i.e. a sample size which is as close as possible to the requested sample size) is suggested by the algorithm.

The output consists of the defining contrast(s) and a corresponding design. In case of a fractional design a particular fraction can be selected by excluding some factor level combinations from the design.

5. Examples:

(Main effects are always included in the model)

Numbers of factor levels $p_1, \ldots, p_K$	size of complete design	interactions of interest	minimum sample size of fractional design	minimum block size
$3^5 (= 3, 3, 3, 3, 3)$	243	1×2 2×4	27	9
3^5	243	1×2 4×5	81	9
3^5	243	all 2-factor interactions	81	27
$5^3 (= 5, 5, 5)$	125	—	25	5
5^3	125	1×2	125	25
2,3,3,6,6	648	—	36	18
2,3,4,6,8	1152	—	576	48
2,2,3,3,6,6	1296	—	72	36
2,2,3,3,6,6	1296	3×5	216	36

For example 6 (numbers of factor levels 2,3,3,6,6, no interactions of interest), the final output of the algorithm is listed below :


```
=========================================================

Summary :
---------

Number of factors : 5

Factor-levels
  2  3  3  6  6

The model includes  :
1. : all main effects
2. : no interactions

fractional design

defining contrasts :

  0  0  1  2  2

  1  1  0  1  5

factor of reduction : 18

size of design : 36


=========================================================
```

```
Notation :
===============

rows represent factor levels in the following way :
0 : factor on lowest level
1 : factor on second level
etc.
```

A	B	C	D	E		A	B	C	D	E
0	0	0	0	0		1	0	0	0	3
0	0	0	3	3		1	0	0	3	0
0	0	1	1	1		1	0	1	1	4
0	0	1	4	4		1	0	1	4	1
0	0	2	2	2		1	0	2	2	5
0	0	2	5	5		1	0	2	5	2
0	1	0	2	4		1	1	0	2	1
0	1	0	5	1		1	1	0	5	4
0	1	1	0	2		1	1	1	0	5
0	1	1	3	5		1	1	1	3	2
0	1	2	1	3		1	1	2	1	0
0	1	2	4	0		1	1	2	4	3
0	2	0	1	5		1	2	0	1	2
0	2	0	4	2		1	2	0	4	5
0	2	1	2	0		1	2	1	2	3
0	2	1	5	3		1	2	1	5	0
0	2	2	0	4		1	2	2	0	1
0	2	2	3	1		1	2	2	3	4

```
==============================================================
```

6. References

[1] BAILEY, R.A. ,GILCHRIST, F.H.L. ,PATTERSON, H.D. (1977) : Identification of Effects and Confounding Patterns in Factorial Designs. Biometrika 64, 347-354

[2] FRANKLIN, M.F. (1985) : Selecting Defining Contrasts and Confounded Effects in p^{n-m} Factorial Experiments. Technometrics, Vol. 27, No. 2, 165-172

[3] GREENFIELD, A.A. (1976) : Selection of Defining Contrasts in Two-level Experiments. Appl. Statist. 25, No. 1, 64-67

[4] GREENFIELD, A.A. (1978) : Selection of Defining Contrasts in Two-level Experiments - A Modification. Appl. Statist. 27, No. 1, 78

[5] HEDAYAT, A., WALLIS, W.D. (1978) : Hadamard Matrices and their Applications. Ann. Statist., Vol. 6, No. 6, 1184-1238

[6] PLACKETT, R.L., BURMAN, J.P. (1946) : The Design of Optimum Multifactorial Experiments. Biometrika 33, 305-325

[7] TURIEL, T.P (1988) : A FORTRAN Program to Generate Fractional Factorial Experiments. J. Quality Technology, Vol. 20, No. 1, 63-72

[8] RIEBSCHLÄGER, M (1989): Ein Algorithmus zur Konstruktion orthogonaler faktorieller Versuchspläne. (unpublished) Ph. D. thesis, Technische Universität Berlin

[9] ZEMROCH, P.J. (1986) : The Computerized Generation of Blocked Incomplete Factorial Designs in the Conversational Experimental Design and Analysis Package SELINA. COMPSTAT 1986, Short Communications, 227-228

[10] ZEMROCH, P.J. (1988) : Strategies for Generating Blocked Fractional Replicate Designs by Computer. Computational Statistics Quaterly, 1, 43-57